THE SIGNET/HAMMOND

WORLD
ATLAS

W9-AZH-892

PRESENTED BY
TIME

Copyright © 1978, 1982 by Hammond Incorporated

All rights reserved

Library of Congress Cataloging in Publication Data
Main entry under title:

The Signet/Hammond world atlas.

Includes index.
1. Atlases. I. Hammond Incorporated.
G1021.S5 1982   912   81-675584
ISBN 0-451-11577-5   AACR2

NEW AMERICAN LIBRARY

TIMES MIRROR

PRINTED IN THE UNITED STATES OF AMERICA

CONTENTS

GAZETTEER-INDEX OF THE WORLD

This alphabetical list of grand divisions, countries, states, colonial possessions, etc., gives area, population, capital, seat of government or chief town, and index references and numbers of plates on which they are shown on the largest scale. The index reference shows the square on the respective map in which the name of the country, state or colonial possession is located. *Indicates members of the United Nations.

Country	Area (Sq. Miles)	Population	Capital or Chief Town	Index Ref.	Plate No.
*Afghanistan	250,775	15,540,000	Kabul	B 1	64
Africa	11,707,000	469,000,000			76-77
Alabama, U.S.A.	51,609	3,890,061	Montgomery		114-115
Alaska, U.S.A.	586,412	400,481	Juneau		116-117
*Albania	11,100	2,590,600	Tiranë	B 3	49
Alberta, Canada	255,285	2,153,200	Edmonton		230-231
*Algeria	919,591	17,422,000	Algiers	E 3	78
American Samoa	76	32,395	Pago Pago	K 7	85
Andorra	188	31,000	Andorra la Vella	G 1	41
*Angola	481,351	6,761,000	Luanda	D 4	82
Anguilla	35	6,519	The Valley	F 3	45
Antarctica	5,500,000				90-91
Antigua & Barbuda	171	72,000	St. John's	G 3	107
*Argentina	1,072,072	27,862,771	Buenos Aires		98-99
Arizona, U.S.A.	113,909	2,717,866	Phoenix		118-119
Arkansas, U.S.A.	53,104	2,285,513	Little Rock		120-121
Armenian S.S.R., U.S.S.R.	11,506	3,031,000	Erivan	F 6	51
Asia	17,128,500	2,633,000,000			54-55
*Australia	2,967,909	14,727,000	Canberra		86-89
*Austria	32,375	7,507,000	Vienna		46
Azerbaidzhan S.S.R., U.S.S.R.	33,436	6,028,000	Baku	G 6	51
*Bahamas	5,382	223,455	Nassau	C 1	106
*Bahrain	240	364,000	Manama	G 4	57
*Bangladesh	55,126	87,052,024	Dacca	E 2	65
*Barbados	166	249,000	Bridgetown	G 4	107
*Belgium	11,781	9,855,110	Brussels		37
*Belize	8,867	144,857	Belmopan	B 1	104
*Benin	43,483	3,338,240	Porto-Novo	E 7	79
Bermuda	21	67,651	Hamilton	G 2	107
*Bhutan	18,147	1,240,000	Thimphu	E 2	65
*Bolivia	424,163	5,600,000	La Paz, Sucre	D 7	95
*Botswana	224,764	819,000	Gaborone	E 6	82
*Brazil	3,286,470	119,024,600	Brasília		94-95
British Columbia, Can.	366,253	2,701,900	Victoria		232-233
British Indian Ocean Territory	29		(London, U.K.)	F 6	55
British Virgin Islands	59	12,000	Road Town	H 1	107
Brunei	2,226	213,000	Bandar Seri Begawan	E 5	74
*Bulgaria	42,823	8,862,000	Sofia	D 3	48
*Burma	261,789	32,913,000	Rangoon	A 2	66
*Burundi	10,747	4,021,910	Bujumbura	F 2	83
*Byelorussian (White-Russian) S.S.R., U.S.S.R.	80,154	9,560,000	Minsk	C 4	51
California, U.S.A.	158,693	23,668,562	Sacramento		122-123
*Cambodia (Kampuchea)	69,898	5,200,000	Phnom Penh	D 4	67
*Cameroon	183,568	8,503,000	Yaoundé	G 7	79

4

Country	Area (Sq. Miles)	Population	Capital or Chief Town	Index Ref.	Plate No.
* Canada	3,851,809	24,150,000	Ottawa		108-109
* Cape Verde	1,557	314,000	Praia	G 9	77
Cayman Islands	100	16,677	Georgetown	B 3	106
* Central African Republic	242,000	2,370,000	Bangui	A 6	81
Central America	197,575	22,700,000			104-105
* Ceylon (Sri Lanka)	25,332	14,859,295	Colombo	D 4	64
* Chad	495,752	4,309,000	N'Djamena	A 4	81
Channel Islands	74	133,000	Saint Helier, St. Peter Port.	E 6	31
* Chile	292,257	11,198,789	Santiago		98-99
* China (People's Rep.)	3,691,000	958,090,000	Peking		68-69
China (Taiwan)	13,971	16,609,961	Taipei	F 3	69
* Colombia	439,513	27,520,000	Bogotá	C 3	94
Colorado, U.S.A.	104,247	2,888,834	Denver		124-125
* Comoros	719	290,000	Moroni	J 4	83
* Congo	132,046	1,459,000	Brazzaville	C 2	82
Connecticut, U.S.A.	5,009	3,107,576	Hartford		126-127
Cook Islands	91	18,128	Avarua	L 7	85
* Costa Rica	19,575	2,245,000	San José	C 3	104
* Cuba	44,206	9,833,000	Havana	B 2	106
* Cyprus	3,473	629,000	Nicosia	E 5	58
* Czechoslovakia	49,373	15,276,799	Prague		46-47
Delaware, U.S.A.	2,057	595,225	Dover		128-129
* Denmark	16,629	5,124,000	Copenhagen	A 3	33
District of Columbia, U.S.A.	67	637,651	Washington	A 3	152
* Djibouti	8,880	250,000	Djibouti	F 5	81
* Dominica	290	81,500	Roseau	G 4	107
* Dominican Republic	18,704	5,431,000	Santo Domingo	D 3	106
* East Germany (German Democratic Rep.)	41,768	16,737,000	Berlin (East)		34-35
* Ecuador	109,483	8,354,000	Quito	B 4	94
* Egypt	386,659	41,572,000	Cairo	C 2	80
* El Salvador	8,260	4,813,000	San Salvador	B 2	104
England, U.K.	50,516	46,220,955	London		30-31
* Equatorial Guinea	10,831	346,000	Malabo	F 8	79
Estonian S.S.R., U.S.S.R.	17,413	1,466,000	Tallinn	C 3	50
* Ethiopia	471,776	31,065,000	Addis Ababa	E 5	81
Europe	4,057,000	676,000,000			28-29
Faerøe Islands, Den.	540	44,000	Tórshavn	D 2	28
Falkland Islands	6,198	1,812	Stanley	D 8	99
* Fiji	7,055	588,068	Suva	H 7	84
* Finland	130,128	4,788,000	Helsinki		32-33
Florida, U.S.A.	58,560	9,739,992	Tallahassee		130-131
* France	210,038	53,788,000	Paris		38-39
French Guiana	35,135	64,000	Cayenne	B 2	96
French Polynesia	1,544	137,382	Papeete	M 7	85
* Gabon	103,346	538,000	Libreville	C 2	82
* Gambia	4,127	601,000	Banjul	A 6	79
Georgia, U.S.A.	58,876	5,464,265	Atlanta		132-133
Georgian S.S.R., U.S.S.R.	26,911	5,015,000	Tbilisi	F 6	51
* Germany, East (German Democratic Rep.)	41,768	16,737,000	Berlin (East)		34-35
* Germany, West (Federal Republic)	95,985	61,658,000	Bonn		34-35
* Ghana	92,099	11,450,000	Accra	D 7	79
Gibraltar	2.28	29,760	Gibraltar	D 4	40
* Great Britain & Northern Ireland (U.K.)	94,399	55,671,000	London		30-31
* Greece	50,944	9,599,000	Athens		49
Greenland, Denmark	840,000	49,773	Godthåb	P 2	100

5

Country	Area (Sq. Miles)	Population	Capital or Chief Town	Index Ref.	Plate No.
* Grenada	133	110,000	Saint George's	F 4	107
Guadeloupe	687	319,000	Basse–Terre	F 3	107
Guam	212	105,821	Agaña	E 4	84
* Guatemala	42,042	7,262,419	Guatemala	B 2	104
* Guinea	94,925	5,143,284	Conakry	B 6	79
* Guinea-Bissau	13,948	777,214	Bissau	A 6	79
* Guyana	83,000	824,000	Georgetown	A 1	96
* Haiti	10,694	5,009,000	Port–au–Prince	D 3	106
Hawaii, U.S.A.	6,450	965,000	Honolulu		134-135
* Holland (Netherlands)	15,892	14,220,000	The Hague, Amsterdam		36-37
* Honduras	43,277	3,691,000	Tegucigalpa	C 2	104
Hong Kong	403	4,402,990	Victoria	E 3	69
* Hungary	35,919	10,709,536	Budapest		47
* Iceland	39,768	228,785	Reykjavík	C 2	28
Idaho, U.S.A.	83,557	943,935	Boise		136-137
Illinois, U.S.A.	56,400	11,418,461	Springfield		138-139
* India	1,269,339	683,810,051	New Delhi		64-65
Indiana, U.S.A.	36,291	5,490,179	Indianapolis		140-141
* Indonesia	788,430	147,383,075	Djakarta		74-75
Iowa, U.S.A.	56,290	2,913,387	Des Moines		142-143
* Iran	636,293	37,447,000	Tehran		62-63
* Iraq	172,476	12,767,000	Baghdad		62
* Ireland	27,136	3,364,881	Dublin		30-31
Ireland, Northern, U.K.	5,452	1,545,000	Belfast	C 3	30-31
* Israel	7,847	3,878,000	Jerusalem		60-61
* Italy	116,303	57,140,000	Rome		42-43
* Ivory Coast	124,504	7,920,000	Abidjan	C 7	79
* Jamaica	4,244	2,161,000	Kingston	C 3	106
* Japan	145,730	117,057,485	Tokyo		70-71
* Jordan	37,737	2,152,273	Amman		60-61
Kansas, U.S.A.	82,264	2,363,208	Topeka		144-145
Kazakh S.S.R., U.S.S.R.	1,048,300	14,684,000	Alma–Ata	C 5	52
Kentucky, U.S.A.	40,395	3,661,433	Frankfort		146-147
* Kenya	224,960	15,322,000	Nairobi	E 7	87
Kirgiz S.S.R., U.S.S.R.	76,641	3,529,000	Frunze	D 5	62
Korea, North	46,540	17,072,000	P'yŏngyang		70
Korea, South	38,175	37,448,836	Seoul		70
* Kuwait	6,532	1,355,827	Al Kuwait	F 4	56
* Laos	91,428	3,546,000	Vientiane	D 3	66
Latvian S.S.R., U.S.S.R.	24,595	2,521,000	Riga	C 3	50
* Lebanon	4,015	3,012,000	Beirut	F 6	59
* Lesotho	11,720	1,339,000	Maseru	M 7	83
* Liberia	43,000	1,873,000	Monrovia	C 7	79
* Libya	679,358	2,856,000	Tripoli G 3 78,	A 2	80
Liechtenstein	61	25,220	Vaduz	B 1	42
Lithuanian S.S.R., U.S.S.R.	25,174	3,398,000	Vilna	B 3	50
Louisiana, U.S.A.	48,523	4,203,972	Baton Rouge		148-149
* Luxembourg	999	364,000	Luxembourg	H 8	37
Macau	6.2	271,000	Macau	E 3	69
* Madagascar	226,657	8,289,000	Tananarive	K 6	83
Maine, U.S.A.	33,215	1,124,660	Augusta		150-151
* Malawi	45,747	5,968,000	Lilongwe	G 4	83
* Malaysia	128,308	13,435,588	Kuala Lumpur	D 5	74
* Maldives	115	143,046	Male	C 4	64
* Mali	464,873	6,906,000	Bamako	C 6	79
* Malta	122	343,970	Valletta	E 7	43
Man, Isle of	227	64,000	Douglas	D 3	31
Manitoba, Canada	250,999	1,028,800	Winnipeg		226-227

Country	Area (Sq. Miles)	Population	Capital or Chief Town	Index Ref.	Plate No.
Martinique	425	308,000	Fort–de–France	G 4	107
Maryland, U.S.A.	10,577	4,216,446	Annapolis		152-153
Massachusetts, U.S.A.	8,257	5,737,037	Boston		154-155
* Mauritania	419,229	1,544,000	Nouakchott	B 5	79
* Mauritius	790	959,000	Port Louis	H 7	77
Mayotte	144	47,300	Mamoutzou	J 4	83
* Mexico	761,601	67,395,826	Mexico City		102-103
Michigan, U.S.A.	58,216	9,258,344	Lansing		156-157
Midway Islands	1.9	526		J 3	84
Minnesota, U.S.A.	84,068	4,077,148	St. Paul		158-159
Mississippi, U.S.A.	47,716	2,520,638	Jackson		160-161
Missouri, U.S.A.	69,686	4,917,444	Jefferson City		162-163
Moldavian S.S.R., U.S.S.R.	13,012	3,947,000	Kishinev	C 5	51
Monaco	368 acres	25,029	Monaco	G 6	39
* Mongolia	606,163	1,594,800	Ulaanbaatar		68-69
Montana, U.S.A.	147,138	786,690	Helena		164-165
Montserrat	40	12,073	Plymouth	G 3	107
* Morocco	172,414	20,242,000	Rabat	C 2	78
* Mozambique	303,769	12,130,000	Maputo	G 6	83
Namibia	317,827	1,200,000	Windhoek	D 6	82
Nauru	7.7	7,254	Yaren (dist.)	G 6	84
Nebraska, U.S.A.	77,227	1,570,006	Lincoln		166-167
* Nepal	54,663	14,010,000	Kathmandu	D 2	64
* Netherlands	15,892	14,220,000	The Hague, Amsterdam		36-37
Netherlands Antilles	390	246,000	Willemstad	E 4	107
Nevada, U.S.A.	110,540	799,184	Carson City		168-169
New Brunswick, Canada	28,354	709,600	Fredericton		218-219
New Caledonia	7,335	137,000	Nouméa	G 8	84
Newfoundland, Canada	156,185	584,400	St. John's		214-215
New Hampshire, U.S.A.	9,304	920,610	Concord		170-171
New Jersey, U.S.A.	7,836	7,364,158	Trenton		172-173
New Mexico, U.S.A.	121,666	1,299,968	Santa Fe		174-175
New South Wales, Australia	309,433	4,914,300	Sydney	H 6	89
New York, U.S.A.	49,576	17,557,288	Albany		176-177
* New Zealand	103,736	3,167,357	Wellington	L 6	89
* Nicaragua	45,698	2,703,000	Managua	C 2	104
* Niger	489,189	5,098,427	Niamey	F 5	79
* Nigeria	357,000	84,500,000	Lagos	F 7	79
Niue	100	3,843	Alofi	K 7	85
North America	9,363,000	370,000,000			100-101
North Carolina, U.S.A.	52,586	5,874,429	Raleigh		178-179
North Dakota, U.S.A.	70,665	652,695	Bismarck		180-181
Northern Ireland, U.K.	5,452	652,695	Belfast	C 3	30-31
Northern Terr., Aust.	520,280	97,090	Darwin	E 3	86
North Korea	46,540	17,072,000	P'yŏngyang		70
Northwest Territories, Canada	1,304,897	43,000	Yellowknife		236-237
* Norway	125,053	4,092,000	Oslo		32-33
Nova Scotia, Canada	21,425	856,600	Halifax		216-217
Oceania	3,292,000	23,000,000			84-85
Ohio, U.S.A.	41,222	10,797,419	Columbus		182-183
Oklahoma, U.S.A.	69,919	3,025,266	Oklahoma City		184-185
* Oman	120,000	839,000	Muscat	H 5	57
Ontario, Canada	412,580	8,614,200	Toronto		224-225
Oregon, U.S.A.	96,981	2,632,663	Salem		186-187
Pacific Islands, Territory of the	707	133,836	Kolonia (Ponape)	F 5	84
* Pakistan	310,403	83,700,000	Islamabad	B 2	64

7

Country	Area (Sq. Miles)	Population	Capital or Chief Town	Index Ref.	Plate No.
*Panama	29,209	1,830,175	Panamá	D 3	105
*Papua New Guinea	183,540	3,006,799	Port Moresby	E 6	84
*Paraguay	157,047	2,973,000	Asunción	E 2	98
Pennsylvania, U.S.A.	45,333	11,866,728	Harrisburg		188-189
*Peru	496,222	17,780,000	Lima	B 5	95
*Philippines	115,707	47,914,017	Manila		72-73
Pitcairn Islands	18	61	Adamstown	O 8	85
*Poland	120,725	35,815,000	Warsaw		44-45
*Portugal	35,549	9,933,000	Lisbon		40
Prince Edward I., Can.	2,184	124,200	Charlottetown		220-221
Puerto Rico	3,435	3,186,076	San Juan	G 1	107
*Qatar	4,247	250,000	Doha	G 4	57
Québec, Canada	594,860	6,334,700	Québec		222-223
Queensland, Aust.	666,991	2,111,700	Brisbane		88-89
Réunion	969	491,000	Saint-Denis	H 7	77
Rhode Island, U.S.A.	1,214	947,154	Providence		190-191
*Romania	91,699	22,048,305	Bucharest		48
Russian S.F.S.R., U.S.S.R.	6,592,812	137,551,000	Moscow		50-53
*Rwanda	10,169	4,819,317	Kigali	C 2	83
St. Christopher-Nevis	120	44,404	Basseterre	F 3	107
Saint Helena	162	5,147	Jamestown	B 6	77
Saint Lucia	238	115,783	Castries	G 4	107
Saint Pierre & Miquelon	93.5	6,000	Saint-Pierre	B 4	215
*St. Vincent & the Grenadines	150	124,000	Kingstown	G 4	107
San Marino	23.4	19,149	San Marino	D 3	42
*São Tomé e Príncipe	372	83,000	São Tomé	A 1	82
Saskatchewan, Canada	251,699	977,400	Regina		228-229
*Saudi Arabia	829,995	7,866,000	Riyadh, Mecca		56-57
Scotland, U.K.	30,414	5,117,000	Edinburgh		30
*Senegal	76,124	5,508,000	Dakar	B 6	79
*Seychelles	145	63,000	Victoria	E 6	55
*Sierra Leone	27,925	3,470,000	Freetown	B 7	79
*Singapore	226	2,413,945	Singapore	E 6	67
*Solomon Islands	11,500	221,000	Honiara	F 6	84
*Somalia	246,200	3,443,000	Mogadishu	G 6	81
*South Africa	455,318	24,400,000	Cape Town, Pretoria		82-83
South America	6,875,000	245,000			92-93
South Australia, Aust.	380,070	1,261,600	Adelaide		87
South Carolina, U.S.A.	31,055	3,119,208	Columbia		192-193
South Dakota, U.S.A.	77,047	690,178	Pierre		194-195
South Korea	38,175	37,448,836	Seoul		70
South-West Africa (Namibia)	317,827	1,200,000	Windhoek	D 6	82
*Spain	194,881	37,430,000	Madrid		40-41
*Sri Lanka	25,332	14,859,295	Colombo	D 4	64
*Sudan	967,494	18,691,000	Khartoum	C 5	81
*Suriname	55,144	352,000	Paramaribo	A 2	96
*Swaziland	6,705	547,000	Mbabane	G 7	83
*Sweden	173,665	8,320,000	Stockholm		32-33
Switzerland	15,943	6,365,960	Bern	B 1	42
*Syria	71,498	8,797,000	Damascus	G 5	59
Tadzhik S.S.R., U.S.S.R.	55,251	3,801,000	Dushanbe	C 6	52
*Tanzania	363,708	17,527,564	Dar es Salaam	G 3	83
Tasmania, Australia	26,383	402,866	Hobart	H 8	89
Tennessee, U.S.A.	42,244	4,590,750	Nashville		196-197
Texas, U.S.A.	267,339	14,228,383	Austin		198-199
*Thailand	198,455	46,455,000	Bangkok		66-67

Country	Area (Sq. Miles)	Population	Capital or Chief Town	Index Ref.	Plate No.
Tibet, China	471,660		Lhasa	B 2	68
*Togo	21,622	2,472,000	Lomé	E 7	79
Tokelau	3.9	1,575	Fakaofo	J 6	85
Tonga	270	90,128	Nuku'alofa	J 7	85
*Trinidad and Tobago....	1,980	1,067,108	Port–of–Spain	G 5	107
*Tunisia	63,378	6,367,000	Tunis	F 2	78
*Turkey	300,946	45,217,556	Ankara		58–59
Turkmen S.S.R., U.S.S.R.	188,455	2,759,000	Ashkhabad	B 6	52
Turks and Caicos Is.	166	7,436	Cockburn Town (Grand Turk)	D 2	106
Tuvalu	9.78	7,349	Fongafale (Funafuti)	H 6	84
*Uganda	91,076	12,630,076	Kampala	D 7	81
*Ukrainian S.S.R., U.S.S.R.	233,089	49,755,000	Kiev	D 5	51
*Union of Soviet Socialist Republics	8,649,490	262,436,227	Moscow		50–53
*United Arab Emirates	32,278	1,040,275	Abu Dhabi	G 5	57
*United Kingdom	94,399	55,671,000	London		30–31
*United States of America	3,615,123	226,504,825	Washington		110–111
*Upper Volta	105,869	6,554,000	Ouagadougou	D 6	79
*Uruguay	72,172	2,899,000	Montevideo	E 4	98
Utah, U.S.A.	84,916	1,461,037	Salt Lake City		200–201
Uzbek S.S.R., U.S.S.R.	173,591	15,391,000	Tashkent	C 5	52
*Vanuatu	5,700	112,596	Vila	H 7	84
Vatican City	116 acres	728		D 4	42
*Venezuela	352,143	13,913,000	Caracas	D 2	94
Vermont, U.S.A.	9,609	511,456	Montpelier		202–203
Victoria, Australia	87,884	3,746,000	Melbourne	G 7	89
*Vietnam	128,405	52,741,766	Hanoi		66–67
Virginia, U.S.A.	40,817	5,346,279	Richmond		204–205
Virgin Islands, British	59	12,000	Road Town	H 1	107
Virgin Islands, U.S.A.	133	95,214	Charlotte Amalie	H 1	107
Wake Island	2.5	437	Wake Islet	G 4	84
Wales, U.K.	8,017	2,790,462	Cardiff	E 4	31
Wallis and Futuna	106	9,192	Matautu	J 7	85
Washington, U.S.A.	68,192	4,130,163	Olympia		206–207
Western Australia, Australia	975,920	1,169,800	Perth		86–87
Western Sahara	102,702	165,000		B 3	78
*Western Samoa	1,133	151,983	Apia	J 7	85
*West Germany (Federal Republic)	95,985	61,658,000	Bonn		34–35
West Virginia, U.S.A. ..	24,181	1,949,644	Charleston		208–209
*White Russian (Byelorussian) S.S.R., U.S.S.R.	80,154	9,560,000	Minsk	C 4	51
Wisconsin, U.S.A.	56,154	4,705,335	Madison		210–211
World	57,970,000	4,415,000,000			26–27
Wyoming, U.S.A.	97,914	470,816	Cheyenne		212–213
*Yemen Arab Republic ..	77,220	7,080,000	San'a	E 7	56
*Yemen, People's Dem. Rep. of	111,101	1,969,000	Aden	F 7	56
*Yugoslavia	98,766	22,471,000	Belgrade		48
Yukon Territory, Can. ..	207,076	21,400	Whitehorse		234–235
*Zaire	918,962	28,291,000	Kinshasa	B 8	81, E 2 82
*Zambia	290,586	5,679,808	Lusaka		82–83
*Zimbabwe	150,803	7,360,000	Salisbury	F 5	83

9

GLOSSARY OF GEOGRAPHICAL TERMS

A. = Arabic Camb. = Cambodian Ch. = Chinese Dan. = Danish Du. = Dutch
Finn. = Finnish Fr. = French Ger. = German Ice. = Icelandic It. = Italian
Jap. = Japanese Mong. = Mongol Nor. = Norwegian Per. = Persian
Port. = Portuguese Russ. = Russian Sp. = Spanish Sw. = Swedish Turk. = Turkish

Å	Nor., Sw.	Stream
Abajo	Sp.	Lower
Ada, Adasi	Turk.	Island
Altiplano	Sp.	Plateau
Älv, Alf, Elf	Sw.	River
Arrecife	Sp.	Reef
Baai	Du.	Bay
Bahía	Sp.	Bay
Bahr	Arabic	Marsh, Lake, Sea, River
Baia	Port.	Bay
Baie	Fr.	Bay, Gulf
Bañados	Sp.	Marshes
Barra	Sp.	Reef
Belt	Ger.	Strait
Ben	Gaelic	Mountain
Berg	Ger., Du.	Mountain
Bir	Aarbic	Well
Boca	Sp.	Gulf, Inlet
Bolshoi, Bolshaya	Russ.	Big
Bolsón	Sp.	Depression
Bong	Korean	Mountain
Bucht	Ger.	Bay
Bugt	Dan.	Bay
Bukhta	Russ.	Bay
Burnu, Burun	Turk.	Cape, Point
By	Dan., Nor., Sw.	Town
Cabo	Port., Sp.	Cape
Campos	Port.	Plains
Canal	Port., Sp.	Channel
Cap, Capo	Fr., It.	Cape
Catarátas	Sp.	Falls
Central, Centrale	Fr., It.	Middle
Cerrito, Cerro	Sp.	Hill
Ciénaga	Sp.	Swamp
Ciudad	Sp.	City
Col	Fr.	Pass
Cordillera	Sp.	Mt. Range
Côte	Fr.	Coast
Cuchilla	Sp.	Mt. Range
Dağ, Dagh	Turk.	Mountain
Dağlari	Turk.	Mt. Range
Dal	Nor., Sw.	Valley
Darya	Per.	Salt Lake
Dasht	Per.	Desert, Plain
Deniz, Denizi	Turk.	Sea, Lake
Desierto	Sp.	Desert
Eiland	Du.	Island
Elv	Dan., Nor.	River
Emi	Berber	Mountain
Erg	Arabic	Dune, Desert
Est, Este	Fr., Port., Sp.	East
Estrecho, Estreito	Sp., Port.	Strait
Étang	Fr.	Pond, Lagoon, Lake
Fjörd	Dan., Nor.	Fiord
Fleuve	Fr.	River
Gebel	Arabic	Mountain
Gebirge	Ger.	Mt. Range
Gobi	Mongol	Desert
Gol	Mongol, Turk.	Lake, Stream
Golf	Ger., Du.	Gulf
Golfe	Fr.	Gulf
Golfo	Sp., It., Port.	Gulf
Gölü	Turk.	Lake
Gora	Russ.	Mountain
Grand, Grande	Fr., Sp.	Big
Groot	Du.	Big
Gross	Ger.	Big
Grosso	It., Port.	Big
Guba	Russ.	Bay, Gulf
Gunto	Jap.	Archipelago
Gunung	Malay	Mountain
Higashi, Higasi	Jap.	East
Ho	Ch.	River
Hoek	Du.	Cape
Holm	Dan., Nor., Sw.	Island
Hu	Ch.	Lake
Hwang	Ch.	Yellow
Île	Fr.	Island
Insel	Ger.	Island
Irmak	Turk.	River
Isla	Sp.	Island
Isola	Sp.	Island
Jabal, Jebel	Arabic	Mountains
Järvi	Finn.	Lake
Jaure	Sw.	Lake
Jezira	Arabic	Island
Jima	Jap.	Island
Joki	Finn.	River
Kaap	Du.	Cape
Kabir, Kebir	Arabic	Big
Kanal	Russ., Ger.	Canal, Channel
Kap, Kapp	Nor., Sw., Ice.	Cape
Kawa	Jap.	River
Khrebet	Russ.	Mt. Range
Kiang	Ch.	River
Kita	Jap.	North
Klein	Du., Ger.	Small
Kô	Jap.	Lake
Ko	Thai.	Island
Koh	Camb., Khmer	Island
Köping	Sw.	Borough
Körfez, Körfezi	Turk.	Gulf
Kuh	Per.	Mountain

10

Term	Language	Meaning
Kul	Sinkiang Turki	Lake
Kum	Turk.	Desert
Lac	Fr.	Lake
Lago	Port., Sp., It.	Lake
Lagôa	Port.	Lagoon
Laguna	Sp.	Lagoon
Lagune	Fr.	Lagoon
Llanos	Sp.	Plains
Mar	Sp., Port.	Sea
Mare	It.	Sea
Meer	Du.	Lake
Meer	Ger.	Sea
Mer	Fr.	Sea
Meseta	Sp.	Plateau
Minami	Jap.	Southern
Misaki	Jap.	Cape
Mittel	Ger.	Middle
Mont	Fr.	Mountain
Montagne	Fr.	Mountain
Montaña	Sp.	Mountains
Monte	Sp., It., Port.	Mountain
More	Russ.	Sea
Muong	Siamese	Town
Mys	Russ.	Cape
Nam	Burm., Lao	River
Nevado	Sp.	Snow covered peak
Nieder	Ger.	Lower
Nishi, Nisi	Jap.	West
Nizhni, Nizhnyaya	Russ.	Lower
Nor	Mong.	Lake
Nord	Fr., Ger.	North
Norte	Sp., It., Port.	North
Nos	Russ.	Cape
Novi, Novaya	Russ.	New
Nusa	Malay	Island
O	Jap.	Big
Ö	Nor., Sw	Island
Ober	Ger.	Upper
Occidental, Occidentale	Sp., It.	Western
Oeste	Port.	West
Oriental	Sp., Fr.	Eastern
Orientale	It.	Eastern
Ost	Ger.	East
Ostrov	Russ.	Island
Ouest	Fr.	West
Öy	Nor.	Island
Ozero	Russ.	Lake
Pampa	Sp.	Plain
Paso	Sp.	Pass
Passo	It., Port.	Pass
Pequeño	Sp.	Small
Peski	Russ.	Desert
Petit	Fr.	Small
Pic	Fr.	Mountain
Pico	Port., Sp.	Mountain, Peak
Pik	Russ.	Peak
Pointe	Fr.	Point
Poluostrov	Russ.	Peninsula
Ponta	Port.	Point
Presa	Sp.	Reservoir
Proliv	Russ.	Strait
Pulou, Pulo	Malay	Island
Punta	Sp., It., Port.	Point
Ras	Arabic	Cape
Ría	Sp.	Estuary
Río	Sp.	River
Rivier, Rivière	Du., Fr.	River
Rud	Per.	River
Saki	Jap.	Cape
Salto	Sp., Port.	Falls
San	Ch., Jap., Korean	Hill
See	Ger.	Sea, Lake
Selvas	Sp., Port.	Forest
Serra	Port.	Mts.
Serranía	Sp.	Mts.
Severni, Servernaya	Russ.	North
Shan	Ch., Jap.	Hill, Mts.
Shima	Jap.	Island
Shoto	Jap.	Islands
Sierra	Sp.	Mountains
Sjö	Nor., Sw.	Lake, Sea
Spitze	Ger.	Mt. Peak
Sredni, Srednyaya	Russ.	Middle
Stad	Dan., Nor., Sw.	City
Stari, Staraya	Russ.	Old
Su	Turk.	River
Sud, Süd	Sp., Fr., Ger.	South
Sul	Port.	South
Sungei	Malay	River
Sur	Sp.	South
Tagh	Turk.	Mt. Range
Tal	Ger.	Valley
Tandjong, Tanjung	Malay	Cape, Point
Tso	Tibetan	Lake
Val	Fr.	Valley
Velho	Port.	Old
Verkhni	Russ.	Upper
Vesi	Finn.	Lake
Vishni, Vishnyaya	Russ.	High
Vostochni, Vostochnaya	Russ.	East, Eastern
Wadi	Arabic	Dry River
Wald	Ger.	Forest
Wan	Jap.	Bay
Yama	Jap.	Mountain
Yug, Yuzhni, Yuzhnaya	Russ.	South, Southern
Zaliv	Russ.	Bay, Gulf
Zapadni, Zapadnaya	Russ.	Western
Zee	Du.	Sea
Zemlya	Russ.	Land
Zuid	Du.	South

WORLD STATISTICAL TABLES

OCEANS AND SEAS

	AREA IN SQ. MILES	GREATEST DEPTH IN FEET	VOLUME IN CUBIC MILES
Pacific Ocean	64,186,000	36,198	167,025,000
Atlantic Ocean . . .	31,862,000	28,374	77,580,000
Indian Ocean	28,350,000	25,344	68,213,000
Arctic Ocean	5,427,000	17,880	3,026,000
Caribbean Sea	970,000	24,720	2,298,400
Mediterranean Sea . .	969,000	16,896	1,019,400
South China Sea . . .	895,000	15,000	....
Bering Sea	875,000	15,800	788,500
Gulf of Mexico	600,000	12,300	....
Sea of Okhotsk . . .	590,000	11,070	454,700
East China Sea	482,000	9,500	52,700
Japan Sea	389,000	12,280	383,200
Hudson Bay	317,500	846	37,590
North Sea	222,000	2,200	12,890
Black Sea	185,000	7,365	....
Red Sea	169,000	7,200	53,700
Baltic Sea	163,000	1,506	5,360

PRINCIPAL MOUNTAINS

	FEET		FEET
Everest, Nepal-China .	29,028	Logan, Yukon	19,850
K2 (Godwin Austen),		Cotopaxi, Ecuador . . .	19,347
Pakistan	28,250	Kilimanjaro, Tanzania . .	19,340
Kanchenjunga, Nepal-India	28,208	El Misti, Peru	19,101
Lhotse, Nepal-China . .	27,923	Huila, Colombia . . .	18,865
Makalu, Nepal-China . .	27,824	Citlaltepetl (Orizaba),	
Dhaulagiri, Nepal . . .	26,810	Mexico	18,855
Nanga Parbat, Pakistan .	26,660	El'brus, U.S.S.R. . . .	18,510
Annapurna, Nepal . . .	26,504	Demavend, Iran	18,376
Nanda Devi, India . .	25,645	St. Elias, Alaska-Yukon .	18,008
Kamet, India	25,447	Popocatepetl, Mexico . .	17,887
Tirich Mir, Pakistan . .	25,230	Dykh-Tau, U.S.S.R. . . .	17,070
Minya Konka, China . .	24,902	Kenya, Kenya	17,058
Muztagh Ata, China . .	24,757	Ararat, Turkey	16,946
Communism Peak, U.S.S.R.	24,599	Vinson Massif, Antarc. .	16,864
Pobeda Peak, U.S.S.R. .	24,406	Margherita (Ruwenzori),	
Chomo Lhari, Bhutan-China	23,997	Africa	16,795
Muztagh, China . . .	23,891	Kazbek, U.S.S.R. . . .	16,512
Aconcagua, Argentina . .	22,831	Djaja, Indonesia . . .	16,503
Ojos del Salado, Chile-Arg.	22,572	Blanc, France	15,771
Tupungato, Chile-Arg. .	22,310	Klyuchevskaya Sopka,	
Mercedario, Argentina . .	22,211	U.S.S.R.	15,584
Huascaran, Peru . . .	22,205	Rosa (Dufourspitze), Italy-	
Llullaillaco, Chile-Arg. .	22,057	Switzerland	15,203
Ancohuma, Bolivia . .	21,489	Ras Dashan, Ethiopia . .	15,157
Illampu, Bolivia . . .	21,276	Matterhorn, Switzerland .	14,688
Chimborazo, Ecuador . .	20,561	Whitney, California . . .	14,494
McKinley, Alaska . . .	20,320		

WORLD STATISTICAL TABLES
LAKES AND INLAND SEAS

	AREA IN SQ. MILES		AREA IN SQ. MILES
Caspian Sea	143,243	Lake Chad	5,300
Lake Superior	31,700	Lake Onega	3,710
Lake Victoria	26,724	Lake Titicaca	3,200
Aral Sea	25,676	Lake Nicaragua	3,100
Lake Huron	23,010	Lake Athabasca	3,064
Lake Michigan	22,300	Reindeer Lake	2,568
Lake Tanganyika	12,650	Lake Turkana	2,463
Lake Baykal	12,162	Issyk-Kul'	2,425
Great Bear Lake	12,096	Vanern	2,156
Lake Nyasa	11,555	Lake Winnipegosis	2,075
Great Slave Lake	11,269	Lake Mobutu Sese Seko	2,075
Lake Erie	9,910	Kariba Lake	2,050
Lake Winnipeg	9,417	Lake Urmia	1,815
Lake Ontario	7,340	Lake of the Woods	1,679
Lake Ladoga	7,104	Lake Peipus	1,400
Lake Balkhash	7,027	Great Salt Lake	1,100

LONGEST RIVERS

	LENGTH IN MILES		LENGTH IN MILES
Nile, Africa	4,145	Zambezi, Africa	1,600
Amazon, S.A.	3,915	Nelson, Canada	1,600
Mississippi-Missouri, U.S.A.	3,710	Orinoco, S.A.	1,600
Yangtze, China	3,434	Paraguay, S.A.	1,584
Ob-Irtysh, U.S.S.R.	3,362	Kolyma, U.S.S.R.	1,562
Yenisey-Angara, U.S.S.R.	3,100	Ganges, Asia	1,550
Hwang (Yellow), China	2,877	Ural, U.S.S.R.	1,509
Amur, Asia	2,744	Japura, S.A.	1,500
Lena, U.S.S.R.	2,734	Arkansas, U.S.A.	1,450
Congo, Africa	2,718	Colorado, U.S.A.-Mexico	1,450
Mackenzie-Peace, Canada	2,635	Negro, S.A.	1,400
Mekong, Asia	2,610	Dnieper, U.S.S.R.	1,368
Niger, Africa	2,548	Orange, Africa	1,350
Parana, S.A.	2,450	Irrwaddy, Burma	1,325
Murray-Darling, Australia	2,310	Ohio-Allegheny, U.S.A.	1,306
Volga, U.S.S.R.	2,194	Kama, U.S.S.R.	1,262
Madeira, S.A.	2,013	Columbia, U.S.A.-Canada	1,243
Purus, S.A.	1,995	Red, U.S.A.	1,222
Yukon, Alaska-Canada	1,979	Don, U.S.S.R.	1,222
St. Lawrence, Canada-U.S.	1,900	Brazos, U.S.A.	1,210
Rio Grande, U.S.-Mexico	1,885	Saskatchewan, Canada	1,205
Syr-Dar'ya, U.S.S.R.	1,859	Peace-Finlay, Canada	1,195
Sao Francisco, Brazil	1,811	Tigris, Asia	1,181
Indus, Asia	1,800	Darling, Australia	1,160
Danube, Europe	1,775	Angara, U.S.S.R.	1,135
Salween, Asia	1,770	Sungari, Asia	1,130
Brahmaputra, Asia	1,700	Pechora, U.S.S.R.	1,124
Euphrates, Asia	1,700	Snake, U.S.A.	1,038
Tocantins, Brazil	1,677	Churchill, Canada	1,000
Si, China	1,650	Pilcomayo, S.A.	1,000
Amu-Dar'ya, Asia	1,616	Magdalena, Colombia	1,000

AIR DISTANCES BETWEEN MAJOR WORLD CITIES

SOURCE: USAF Aeronautical Chart and Information Center (in statute miles)

	Bangkok	Berlin	Cairo	Cape Town	Caracas	Chicago	Hong Kong	Honolulu	Istanbul	Lima	London	Madrid	Melbourne
Accra	6,850	3,330	2,672	2,974	4,576	5,837	7,615	10,052	3,039	5,421	3,169	2,412	9,32
Amsterdam	5,707	360	2,015	5,997	4,883	4,118	5,772	7,254	1,372	6,538	222	921	10,28
Anchorage	6,022	4,545	6,116	10,478	5,353	2,858	5,073	2,778	5,388	6,385	4,491	5,181	7,72
Athens	4,930	1,121	671	4,957	5,815	5,447	5,316	8,353	352	7,312	1,488	1,474	9,29
Auckland	4,645	9,995	8,825	6,574	9,620	9,507	4,625	5,346	9,203	7,989	10,570	10,884	1,6
Baghdad	3,756	2,029	798	4,924	7,020	6,430	4,260	8,399	1,006	8,487	2,547	2,675	8,10
Bangkok	—	5,351	4,521	6,301	10,558	8,569	1,076	6,610	4,648	12,241	5,929	6,334	4,57
Beirut	4,272	1,689	341	4,794	6,520	6,097	4,756	8,536	614	7,972	2,151	2,190	8,57
Belgrade	5,073	623	1,147	5,419	5,587	5,000	5,327	7,882	500	7,169	1,053	1,263	9,57
Berlin	5,351	—	1,768	5,958	5,242	4,415	5,443	7,323	1,075	6,893	580	1,162	9,92
Bombay	1,870	3,915	2,717	5,103	9,034	8,066	2,679	8,036	3,000	10,389	4,478	4,689	6,10
Buenos Aires	10,490	7,395	7,360	4,285	3,155	5,582	11,478	7,554	7,608	1,945	6,907	6,236	7,21
Cairo	4,521	1,768	—	4,510	6,337	6,116	5,057	8,818	741	7,725	2,158	2,069	8,70
Cape Town	6,301	5,958	4,510	—	6,361	8,489	7,377	11,534	5,204	6,074	5,988	5,306	6,42
Caracas	10,558	5,242	6,337	6,361	—	2,500	10,171	6,024	6,050	1,699	4,662	4,351	9,70
Chicago	8,569	4,415	6,116	8,489	2,500	—	7,797	4,256	5,485	3,772	3,960	4,192	9,66
Copenhagen	5,361	222	1,964	6,179	5,215	4,263	5,392	7,101	1,252	6,886	595	1,289	9,93
Denver	8,409	5,092	6,846	9,331	3,078	920	7,476	3,346	6,164	3,986	4,701	5,028	8,75
Frankfurt (W. Germany)	5,581	270	1,817	5,815	5,022	4,344	5,709	7,450	1,160	6,660	408	885	10,13
Helsinki	4,903	689	2,069	6,490	5,834	4,442	4,867	6,818	1,330	7,349	1,135	1,835	9,44
Hong Kong	1,076	5,443	5,057	7,377	10,171	7,797	—	5,557	4,989	11,415	5,986	6,556	4,60
Honolulu	6,610	7,323	8,818	11,534	6,024	4,256	5,557	—	8,118	5,944	7,241	7,874	5,50
Houston	9,261	5,337	7,005	8,608	2,262	942	8,349	3,902	6,400	3,123	4,860	5,014	8,97
Istanbul	4,648	1,075	741	5,204	6,050	5,485	4,989	8,118	—	7,593	1,551	1,701	9,10
Karachi	2,305	3,365	2,222	5,153	8,502	7,564	2,977	8,059	2,457	9,943	3,928	4,152	6,64
Keflavik	6,300	1,505	3,267	7,107	4,269	2,942	6,044	6,085	2,578	5,965	1,188	1,802	10,55
Kinshasa	5,974	3,916	2,618	2,047	5,752	7,085	6,904	11,178	3,241	6,322	3,951	3,305	8,11
Leningrad	4,718	826	2,034	6,500	5,843	4,589	4,687	6,816	1,306	7,534	1,307	1,985	9,26
Lima	12,241	6,893	7,725	6,074	1,699	3,772	11,415	5,944	7,593	—	6,316	5,907	8,05
Lisbon	6,651	1,442	2,352	5,301	4,040	4,001	6,862	7,835	2,015	5,591	989	317	11,04
London	5,929	580	2,158	5,988	4,662	3,960	5,895	7,241	1,551	6,316	—	786	10,50
Madrid	6,334	1,162	2,069	5,306	4,351	4,192	6,556	7,874	1,701	5,907	786	—	10,76
Melbourne	4,579	9,929	8,700	6,428	9,703	9,667	4,605	5,501	9,100	8,052	10,508	10,766	—
Mexico City	9,793	6,054	7,677	8,516	2,234	1,688	8,789	3,791	7,106	2,635	5,558	5,642	8,42
Montreal	8,337	3,740	5,403	7,920	2,443	746	7,736	4,919	4,798	3,967	3,256	3,449	10,39
Moscow	4,394	1,001	1,770	6,277	6,176	4,984	4,443	7,049	1,087	7,855	1,556	2,140	8,96
Nairobi	4,481	3,947	2,217	2,543	7,179	8,012	5,447	10,740	2,957	7,821	4,229	3,840	7,15
New Delhi	1,812	3,598	2,752	5,769	8,837	7,486	2,339	7,413	2,837	10,430	4,178	4,528	6,34
New York City	8,669	3,980	5,598	7,801	2,124	714	8,061	4,969	5,022	3,635	3,473	3,596	10,35
Oslo	5,395	523	2,243	6,477	5,167	4,050	5,342	6,801	1,518	6,857	718	1,485	9,93
Panama City	10,871	5,856	7,118	7,021	867	2,321	10,089	5,254	6,756	1,454	5,285	5,081	9,02
Paris	5,877	549	1,973	5,782	4,735	4,145	5,992	7,452	1,400	6,367	215	652	10,44
Peking	2,027	4,600	4,687	8,034	8,978	6,625	1,195	5,084	4,407	10,365	5,089	5,759	5,63
Rabat	6,652	1,623	2,230	4,954	4,311	4,282	6,954	8,177	2,008	5,590	1,254	474	10,85
Rio de Janeiro	9,987	6,207	6,153	3,773	2,805	5,288	11,002	8,295	6,378	2,351	5,751	5,045	8,21
Rome	5,493	735	1,305	5,231	5,198	4,823	5,773	8,040	853	6,748	892	849	9,94
Saigon (Ho Chi Minh City)	467	5,771	4,987	6,534	10,905	8,695	938	6,302	5,102	12,180	6,345	6,779	4,16
San Francisco	7,930	5,673	7,436	10,248	3,908	1,860	6,904	2,397	6,711	4,516	5,369	5,806	7,85
Santiago	10,967	7,772	7,967	4,947	3,033	5,295	11,615	6,861	8,135	1,528	7,241	6,639	7,01
Seattle	7,455	5,060	6,809	10,205	4,099	1,737	6,481	2,681	6,077	4,461	4,799	5,303	8,17
Shanghai	1,797	5,231	5,188	8,062	9,508	7,071	760	4,947	4,975	10,665	5,728	6,386	4,99
Shannon	6,256	940	2,534	6,188	4,320	3,583	6,246	7,006	1,938	5,992	387	884	10,82
Singapore	887	6,167	5,143	6,007	11,403	9,376	1,608	6,728	5,379	11,689	6,747	7,079	3,76
St. Louis	8,763	4,676	6,370	8,549	2,414	265	7,949	4,134	5,744	3,589	4,215	4,426	9,47
Stockholm	5,141	505	2,084	6,422	5,422	4,288	5,115	6,873	1,347	7,109	892	1,613	9,69
Tehran	3,392	2,184	1,220	5,240	7,322	6,502	3,844	8,072	1,274	8,850	2,739	2,974	7,83
Tokyo	2,865	5,557	5,937	9,155	8,813	6,313	1,792	3,860	5,743	9,628	5,956	6,704	5,07
Vienna	5,252	323	1,455	5,656	5,374	4,696	5,432	7,632	791	6,990	767	1,124	9,80
Warsaw	5,032	322	1,588	5,934	5,563	4,679	5,147	7,368	858	7,212	901	1,425	9,60
Washington D.C.	8,807	4,182	5,800	7,892	2,051	2,598	8,157	4,839	5,225	3,504	3,676	3,794	10,17

Mexico City	Montreal	Moscow	Nairobi	New Delhi	New York	Paris	Peking	Rio de Janeiro	Rome	San Francisco	Singapore	Stockholm	Tehran	Tokyo	Vienna	Warsaw
677	5,146	4,038	2,603	5,279	5,126	2,988	7,359	3,501	2,624	7,688	7,183	3,835	3,874	8,594	3,100	3,440
735	3,426	1,337	4,136	3,958	3,654	271	4,890	5,938	807	5,465	6,526	701	2,533	5,788	581	681
776	3,133	4,364	8,287	5,709	3,373	4,697	3,997	8,145	5,263	2,205	6,678	4,102	5,654	3,463	4,856	4,601
021	4,737	1,387	2,827	3,120	4,938	1,305	4,757	6,030	654	6,792	5,629	1,498	1,539	5,924	801	996
274	10,231	9,018	7,315	6,420	10,194	10,519	5,626	8,259	10,048	7,692	3,848	9,732	7,935	5,017	9,886	9,676
082	5,768	1,583	2,431	1,966	6,007	2,405	3,925	6,938	1,836	7,466	4,427	2,164	431	5,199	1,781	1,752
793	8,337	4,394	4,481	1,812	8,669	5,877	2,027	9,987	5,493	7,930	887	5,141	3,392	2,865	5,252	5,032
707	5,405	1,514	2,420	2,479	5,622	1,987	4,352	6,478	1,368	7,302	4,935	1,931	913	5,598	1,401	1,459
610	4,305	1,066	3,328	3,270	4,526	902	4,634	6,145	449	6,296	5,833	1,010	1,741	5,720	309	516
054	3,740	1,001	3,947	3,598	3,980	549	4,600	6,207	735	5,673	6,167	505	2,184	5,557	323	322
739	7,524	3,132	2,811	722	7,811	4,367	2,953	8,334	3,846	8,406	2,427	3,880	1,743	4,196	3,725	3,601
580	5,597	8,369	6,479	9,823	5,279	6,857	11,994	1,231	6,925	6,455	9,870	7,799	8,565	11,411	7,334	7,656
677	5,403	1,770	2,217	2,752	5,598	1,973	4,687	6,153	1,305	7,436	5,143	2,084	1,220	5,937	1,455	1,588
516	7,920	6,277	2,543	5,769	7,801	5,782	8,034	3,773	5,231	10,248	6,007	6,422	5,240	9,155	5,656	5,934
234	2,443	6,176	7,179	8,837	2,124	4,735	8,978	2,805	5,198	3,908	11,408	5,422	7,322	8,813	5,374	5,563
688	746	4,984	8,012	7,486	714	4,145	6,625	5,288	4,823	1,860	9,374	4,288	6,502	6,313	4,696	4,679
918	3,605	971	4,156	3,640	3,857	642	4,503	6,321	953	5,473	6,195	325	2,287	5,415	540	417
438	1,639	5,501	8,867	7,730	1,631	4,900	6,385	5,866	5,887	953	9,079	4,879	7,033	5,815	5,395	5,322
945	3,650	1,261	3,865	3,811	3,857	295	4,853	5,932	595	5,700	6,386	745	2,346	5,831	388	559
101	3,845	554	4,282	3,247	4,126	1,192	3,956	6,872	1,370	5,435	5,759	248	2,062	4,872	895	569
789	7,736	4,443	5,447	2,339	8,061	5,992	1,195	11,002	5,773	6,904	1,608	5,115	3,844	1,792	5,432	5,147
791	4,919	7,049	10,740	7,413	4,969	7,452	5,084	8,295	8,040	2,397	6,728	6,873	8,072	3,860	7,632	7,368
749	1,605	5,925	8,746	8,388	1,419	5,035	7,244	5,015	5,702	1,648	9,954	5,227	7,442	6,685	5,609	5,609
106	4,798	1,087	2,957	2,837	5,022	1,400	4,407	6,308	853	6,711	5,379	1,347	1,274	5,574	791	858
249	6,997	2,600	2,708	678	7,277	3,817	3,020	8,082	3,306	8,078	2,942	3,340	1,194	4,313	3,175	3,052
614	2,317	2,083	5,404	4,749	2,597	1,402	4,951	6,090	2,068	4,196	7,181	1,352	3,568	5,497	1,813	1,745
915	6,378	4,328	4,234	4,692	6,378	3,742	7,002	4,105	3,186	8,920	6,132	4,388	3,612	8,307	3,619	3,910
276	4,005	396	4,230	3,069	4,291	1,350	3,789	7,028	1,460	5,523	5,575	431	1,926	4,733	986	642
635	3,967	7,855	7,821	10,430	3,635	6,367	10,365	2,351	6,748	4,516	11,689	7,109	8,850	9,628	6,990	7,212
396	3,255	2,433	4,013	4,844	3,377	904	6,040	4,777	1,163	5,679	7,393	1,862	3,288	6,943	1,432	1,720
558	3,256	1,556	4,229	4,178	3,473	215	5,089	5,751	892	5,369	6,747	892	2,739	5,956	767	901
642	3,449	2,140	3,840	4,528	3,596	652	5,759	5,045	849	5,806	7,019	1,613	2,974	6,704	1,124	1,425
420	10,390	8,965	7,159	6,340	10,352	10,442	5,723	7,772	9,940	7,850	3,767	9,693	7,838	5,070	9,802	9,609
315	—	4,397	7,267	7,012	333	3,432	6,541	5,082	4,102	2,544	9,207	3,667	5,879	6,470	4,007	4,021
671	4,397	—	3,928	2,703	4,680	1,550	3,627	7,162	1,477	5,884	5,236	764	1,534	4,663	1,039	716
218	7,267	3,928	—	3,371	7,365	4,020	5,720	5,556	3,340	9,598	4,636	4,299	2,709	6,996	3,625	3,800
119	7,012	2,703	3,317	—	7,319	4,103	2,350	8,747	3,684	7,691	2,574	3,466	1,584	3,638	3,467	3,277
086	333	4,680	7,365	7,319	—	3,638	6,867	4,805	4,293	2,574	9,539	3,939	6,141	6,757	4,233	4,271
722	3,418	1,024	4,446	3,726	3,686	838	4,395	6,462	1,248	5,196	6,289	260	2,462	5,238	839	661
496	2,542	6,720	8,043	9,422	2,213	5,388	8,939	3,296	5,916	3,326	11,692	5,956	8,011	8,441	6,031	6,175
723	3,432	1,550	4,020	4,103	3,638	—	5,138	5,681	688	5,579	6,676	964	2,624	6,054	643	853
772	6,541	3,627	5,720	2,350	6,867	5,138	—	10,778	5,076	5,934	2,734	4,197	3,496	1,305	4,664	4,340
612	3,537	2,579	3,733	4,841	3,636	1,125	6,206	4,589	1,184	5,995	7,348	2,084	3,263	7,174	1,546	1,866
769	5,082	7,162	5,556	8,747	4,805	5,681	10,778	—	5,704	6,621	9,776	6,638	7,368	11,535	6,124	6,453
374	4,102	1,477	3,340	3,684	4,293	688	5,076	5,704	—	6,259	6,231	1,229	2,126	6,140	476	819
889	8,558	4,798	4,874	2,268	8,889	6,303	2,072	10,290	5,943	7,829	682	5,534	3,851	2,689	5,687	5,454
889	2,544	5,884	9,598	7,691	2,574	5,579	5,934	6,621	6,259	—	9,449	5,372	7,362	5,148	5,992	5,854
094	5,835	8,770	7,180	10,518	5,106	7,224	11,859	1,820	7,391	5,926	10,190	8,120	9,185	10,711	7,760	8,059
340	2,289	5,217	9,006	7,046	2,409	5,012	5,432	6,890	5,680	679	8,074	4,731	6,686	4,793	5,381	5,222
033	7,067	4,248	5,951	2,646	7,384	5,772	645	11,339	5,679	6,150	2,363	4,837	3,974	1,097	5,281	4,963
172	2,873	1,863	4,563	4,529	3,086	563	5,288	5,597	1,247	5,040	7,089	1,135	3,117	6,064	1,153	1,258
331	9,207	5,236	4,636	2,574	9,539	6,676	2,754	9,776	6,231	8,449	—	5,993	4,106	3,304	6,039	5,846
425	5,629	5,248	8,231	7,736	878	4,398	6,752	5,218	5,073	1,744	5,944	4,552	6,766	6,407	4,955	4,942
965	3,667	764	4,299	3,466	3,939	964	4,197	6,638	1,229	5,372	5,993	—	2,217	5,091	771	504
182	5,879	1,534	2,709	1,584	6,141	2,624	3,496	7,386	2,126	7,362	4,106	2,217	—	4,775	1,983	1,878
036	6,916	4,628	6,996	3,638	6,757	6,054	1,305	11,535	6,148	5,148	3,304	5,091	4,775	—	5,689	5,346
316	4,007	1,039	3,625	3,467	4,233	643	4,664	6,124	476	5,992	6,039	771	1,983	5,689	—	347
335	4,021	716	3,800	3,277	4,271	853	4,340	6,453	819	5,854	5,846	504	1,878	5,346	347	—
883	490	4,873	7,550	7,500	203	3,841	6,965	4,783	4,496	2,444	9,667	4,135	6,340	6,792	4,436	4,471

U.S. AIR DISTANCES

	Atlanta	Birmingham	Boston	Buffalo	Charleston, S.C.	Chicago	Cincinnati	Cleveland	Dallas	Denver	Des Moines	Detroit	Houston	Indianapolis	Jacksonville	Kansas City, Mo.	Los Angeles	Louisville	Miami	Minneapolis	Nashville	New Orleans	New York City	Omaha	Philadelphia	Phoenix	Pittsburgh	Portland, Ore.	St. Louis	Salt Lake City	San Francisco	Seattle	Tulsa	Washington, D.C.
Albuquerque	1272	1095	1972	1580	1539	1129	1229	1421	588	334	837	1364	754	1169	1488	720	664	1178	1690	983	1119	1029	1815	721	1753	330	1580	1181	942	846	896	1184	604	1653
Amarillo																																		
Atlanta	140	140	946	697	267	587	369	554	721	1212	739	596	701	426	303	676	1936	321	604	907	214	424	748	817	666	1456	521	2172	485	1589	2139	2182	678	543
Billings																																		
Birmingham	140		1052	760	361	587	454	618	581	1094	616	641	567	433	374	579	1802	331	699	862	182	312	864	732	780	1456	608	2099	435	1572	2139	2182	581	661
Boston	946	1052		400	851	851	740	551	1551	1769	1159	631	1605	807	1019	1251	2596	826	1255	1123	941	1359	188	1282	271	2300	483	2537	1038	2099	2699	2485	1379	393
Buffalo	697	760	400		699	454	304	186	1208	1370	760	216	1286	435	820	852	2177	483	1181	731	641	1086	292	883	279	2117	186	2198	699	1699	2300	2117	1039	292
Burlington, Vt.																																		
Charleston, S.C.	267	361	851	699		758	498	587	981	1539	947	646	911	583	231	921	2155	507	482	1118	452	681	641	1043	562	1907	539	2534	741	1837	2425	2485	858	453
Charlotte																																		
Cheyenne																																		
Chicago	587	587	851	454	758		252	308	803	920	299	238	940	165	864	405	1745	269	1189	355	397	833	713	416	666	1453	410	1758	262	1260	1858	1737	585	597
Cincinnati	369	454	740	304	498	252		221	814	1094	510	230	875	99	652	533	1829	83	948	595	208	703	589	643	482	1657	257	1972	308	1438	2005	1886	638	400
Cleveland	554	618	551	186	587	308	221		1025	1196	623	90	1105	263	739	700	2026	308	1079	622	429	924	405	744	360	1749	105	2065	492	1566	2166	2026	1025	306
Dallas	721	581	1551	1208	981	803	814	1025		641	569	984	224	763	940	451	1240	731	1111	850	614	443	1374	586	1301	887	1065	1633	547	995	1470	1681	237	1185
Denver	1212	1094	1769	1370	1539	920	1094	1196	641		586	1135	865	996	1467	533	831	1038	1708	693	1011	1038	1631	488	1557	586	1302	986	796	371	949	1021	484	1494
Des Moines	739	616	1159	760	947	299	510	623	569	586		536	879	411	863	166	1479	474	1333	232	576	834	1018	112	983	1100	699	1505	279	983	1550	1467	376	963
Detroit	596	641	631	216	646	238	230	90	984	1135	536		1079	240	831	641	1979	316	1148	528	450	939	486	651	443	1690	205	2000	452	1480	2079	1938	959	396
El Paso																																		
Fargo																																		
Houston	701	567	1605	1286	911	940	875	1105	224	865	879	1079		865	817	643	1374	803	968	1056	667	318	1420	793	1340	1009	1286	1834	679	1195	1645	1891	429	1220
Indianapolis	426	433	807	435	583	165	99	263	763	996	411	240	865		646	453	1797	107	1024	511	249	712	646	525	585	1586	330	1887	229	1356	1949	1872	542	494
Jacksonville	303	374	1019	820	231	864	652	739	940	1467	863	831	817	646		887	2147	594	328	1259	490	465	831	1009	703	1807	658	2445	811	1759	2340	2439	893	589
Kansas City, Mo.	676	579	1251	852	921	405	533	700	451	533	166	641	643	453	887		1360	480	1240	410	491	690	1092	166	1038	1044	759	1491	237	920	1488	1489	219	945
Knoxville																																		
Little Rock																																		
Los Angeles	1936	1802	2596	2177	2155	1745	1829	2026	1240	831	1479	1979	1374	1797	2147	1360		1829	2339	1535	1780	1673	2451	1315	2394	357	2136	834	1589	579	347	959	1256	2300
Louisville	321	331	826	483	507	269	83	308	731	1038	474	316	803	107	594	480	1829		919	605	151	623	652	580	582	1673	344	1950	242	1485	2001	1938	565	476
Memphis																																		
Miami	604	699	1255	1181	482	1189	948	1079	1111	1708	1333	1148	968	1024	328	1240	2339	919		1511	817	669	1092	1397	1019	1982	1010	2708	1061	2089	2594	2734	1176	923
Minneapolis	907	862	1123	731	1118	355	595	622	850	693	232	528	1056	511	1259	410	1535	605	1511		697	1056	1020	281	985	1280	685	1426	466	987	1584	1395	626	934
Nashville	214	182	941	641	452	397	208	429	614	1011	576	450	667	249	490	491	1780	151	817	697		471	761	645	685	1446	472	1969	297	1464	1963	1975	591	562
New Orleans	424	312	1359	1086	681	833	703	924	443	1038	834	939	318	712	465	690	1673	623	669	1056	471		1171	836	1089	1316	919	2101	598	1434	1926	2101	515	966
New York City	748	864	188	292	641	713	589	405	1374	1631	1018	486	1420	646	831	1092	2451	652	1092	1020	761	1171		1144	83	2145	317	2445	875	1972	2571	2408	1266	205
Omaha	817	732	1282	883	1043	416	643	744	586	488	112	651	793	525	1009	166	1315	580	1397	281	645	836	1144		1094	1036	847	1369	347	833	1429	1371	352	1014
Philadelphia	666	780	271	279	562	666	482	360	1301	1557	983	443	1340	585	703	1038	2394	582	1019	985	685	1089	83	1094		2083	259	2412	811	1925	2523	2380	1163	123
Phoenix	1456	1456	2300	2117	1907	1453	1657	1749	887	586	1100	1690	1009	1586	1807	1044	357	1673	1982	1280	1446	1316	2145	1036	2083		1828	1009	1272	504	653	1114	862	1828
Pittsburgh	521	608	483	186	539	410	257	105	1065	1302	699	205	1286	330	658	759	2136	344	1010	685	472	919	317	847	259	1828		2165	559	1723	2412	2165	1020	192
Portland, Ore.	2172	2099	2537	2198	2534	1758	1972	2065	1633	986	1505	2000	1834	1887	2445	1491	834	1950	2708	1426	1969	2101	2445	1369	2412	1009	2165		1708	630	534	145	1486	2354
Raleigh																																		
St. Louis	485	435	1038	699	741	262	308	492	547	796	279	452	679	229	811	237	1589	242	1061	466	297	598	875	347	811	1272	559	1708		1162	1724	1724	361	712
Salt Lake City	1589	1572	2099	1699	1837	1260	1438	1566	995	371	983	1480	1195	1356	1759	920	579	1485	2089	987	1464	1434	1972	833	1925	504	1723	630	1162		601	701	701	917	1848
San Antonio																																		
San Francisco	2139	2139	2699	2300	2425	1858	2005	2166	1470	949	1550	2079	1645	1949	2340	1488	347	2001	2594	1584	1963	1926	2571	1429	2523	653	2412	534	1724	601		678	1490	2442
Seattle	2182	2182	2485	2117	2485	1737	1886	2026	1681	1021	1467	1938	1891	1872	2439	1489	959	1938	2734	1395	1975	2101	2408	1371	2380	1114	2165	145	1724	701	678		1531	2329
Spokane																																		
Syracuse																																		
Tulsa	678	581	1379	1039	858	585	638	1025	237	484	376	959	429	542	893	219	1256	565	1176	626	591	515	1266	352	1163	862	1020	1486	361	917	1490	1531		1058
Washington, D.C.	543	661	393	292	453	597	400	306	1185	1494	963	396	1220	494	589	945	2300	476	923	934	562	966	205	1014	123	1828	192	2354	712	1848	2442	2329	1058	
Wichita	776	658	1424	1036	1039																													

16

WORLD FLAGS and MAPS

AFGHANISTAN

ALBANIA

ALGERIA

ANDORRA

ANGOLA

ANTIGUA & BARBUDA

ARGENTINA

AUSTRALIA

AUSTRIA

BAHAMAS

BAHRAIN

BANGLADESH

BARBADOS

BELGIUM

BELIZE

BENIN (DAHOMEY)

BHUTAN

BOLIVIA

BOTSWANA

BRAZIL

BULGARIA

BURMA

BURUNDI

CAMBODIA (KAMPUCHEA)

17

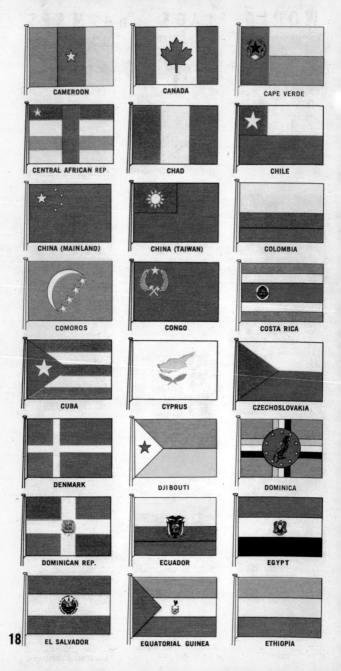

CAMEROON

CANADA

CAPE VERDE

CENTRAL AFRICAN REP.

CHAD

CHILE

CHINA (MAINLAND)

CHINA (TAIWAN)

COLOMBIA

COMOROS

CONGO

COSTA RICA

CUBA

CYPRUS

CZECHOSLOVAKIA

DENMARK

DJIBOUTI

DOMINICA

DOMINICAN REP.

ECUADOR

EGYPT

EL SALVADOR

EQUATORIAL GUINEA

ETHIOPIA

18

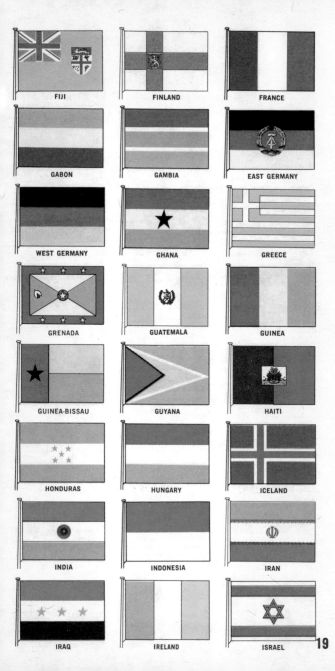

FIJI	FINLAND	FRANCE
GABON	GAMBIA	EAST GERMANY
WEST GERMANY	GHANA	GREECE
GRENADA	GUATEMALA	GUINEA
GUINEA-BISSAU	GUYANA	HAITI
HONDURAS	HUNGARY	ICELAND
INDIA	INDONESIA	IRAN
IRAQ	IRELAND	ISRAEL

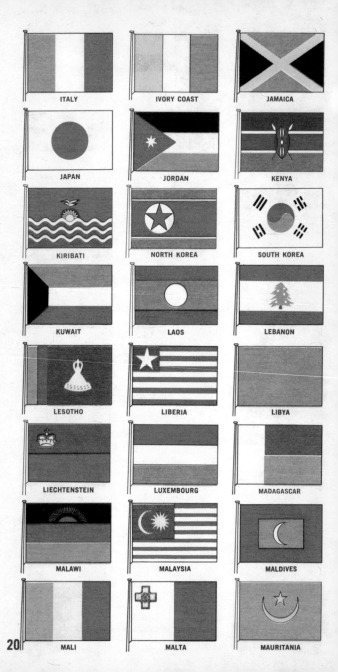

ITALY

IVORY COAST

JAMAICA

JAPAN

JORDAN

KENYA

KIRIBATI

NORTH KOREA

SOUTH KOREA

KUWAIT

LAOS

LEBANON

LESOTHO

LIBERIA

LIBYA

LIECHTENSTEIN

LUXEMBOURG

MADAGASCAR

MALAWI

MALAYSIA

MALDIVES

MALI

MALTA

MAURITANIA

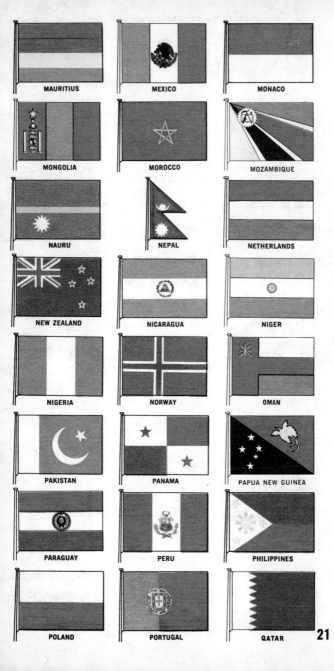

MAURITIUS

MEXICO

MONACO

MONGOLIA

MOROCCO

MOZAMBIQUE

NAURU

NEPAL

NETHERLANDS

NEW ZEALAND

NICARAGUA

NIGER

NIGERIA

NORWAY

OMAN

PAKISTAN

PANAMA

PAPUA NEW GUINEA

PARAGUAY

PERU

PHILIPPINES

POLAND

PORTUGAL

QATAR

21

ROMANIA

RWANDA

ST. LUCIA

ST. VINCENT & GRENS.

SAN MARINO

SÃO TOMÉ E PRÍNCIPE

SAUDI ARABIA

SENEGAL

SEYCHELLES

SIERRA LEONE

SINGAPORE

SOLOMON ISLANDS

SOMALIA

SOUTH AFRICA

SPAIN

SRI LANKA (CEYLON)

SUDAN

SURINAME

SWAZILAND

SWEDEN

SWITZERLAND

SYRIA

TANZANIA

THAILAND

22

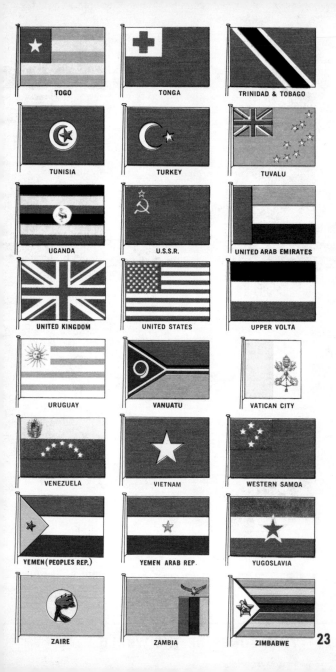

TOGO	TONGA	TRINIDAD & TOBAGO
TUNISIA	TURKEY	TUVALU
UGANDA	U.S.S.R.	UNITED ARAB EMIRATES
UNITED KINGDOM	UNITED STATES	UPPER VOLTA
URUGUAY	VANUATU	VATICAN CITY
VENEZUELA	VIETNAM	WESTERN SAMOA
YEMEN (PEOPLES REP.)	YEMEN ARAB REP.	YUGOSLAVIA
ZAIRE	ZAMBIA	ZIMBABWE

23

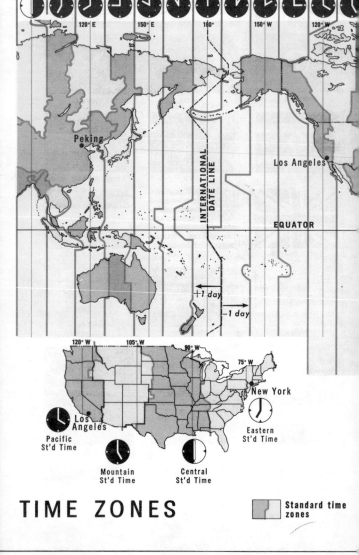

120° E   150° E   180°   150° W   120° W

Peking

Los Angeles

INTERNATIONAL DATE LINE

EQUATOR

←1 day

→1 day

120° W   105° W   90° W   75° W

New York

Los Angeles

Eastern St'd Time

Pacific St'd Time

Mountain St'd Time

Central St'd Time

TIME ZONES

Standard time zones

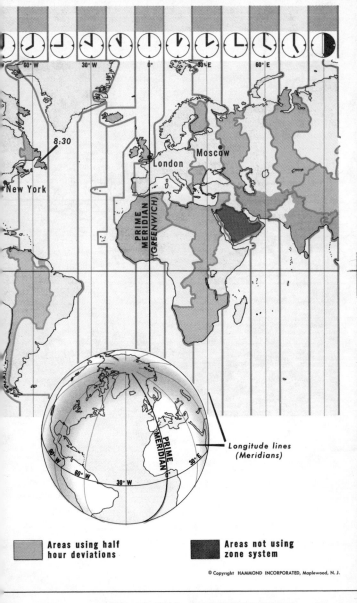

Areas using half hour deviations

Areas not using zone system

TIME ZONES OF THE WORLD — 25

Lincoln Sea
smere land
GREENLAND (KALÂTDLIT-NUNÂT) (Den.)
Thule
BAFFIN BAY
ffin Island
xe sin
Julianehåb
C. Farewell

FRANZ JOSEF LAND (U.S.S.R.)
SVALBARD (Nor.)
GREENLAND SEA
BARENTS SEA
NOVAYA ZEMLYA
KARA SEA
Jan Mayen (Nor.)
North Cape
Murmansk
Dudinka
Salekhard
NORWEGIAN SEA
ICELAND
Reykjavik
Faeroe Is. (Den.)
Arctic Circle
Str. of Denmark
Arctic Circle
Archangel
Ob
Yenisey

Helsinki
UNITED Glasgow North
KINGDOM
IRELAND
London
Stockholm
Baltic
Leningrad
Moscow
Volga
Omsk
UNION OF SOVIET
SOCIALIST REPUBLICS
Novo- sibirsk
Sverdlovsk

tréal
Newfoundland
St. John's
C. Race
Halifax
Boston
New York
Philadelphia
Paris
FRANCE
Warsaw
Khar'kov
Volgograd
Caspian Sea
Aral Sea
Tashkent
Ürümqi

NORTH
PORTUGAL
Lisbon
Madrid
SPAIN
Rome
Mediterranean Sea
Istanbul
TURKEY
Black Sea
Baku
Tehran
Tashkent

Azores (Port.)
Rabat
MOROCCO
Algiers
TUNISIA
Tripoli
ISR.
SYRIA
IRAQ
IRAN
AFGH.
Kabul
Islamabad
New Delhi
SINKIANG
TIBET

ATLANTIC
BAHAMAS
Tropic of Cancer
WEST INDIES
Hispaniola
PUERTO RICO (U.S.)
ribbean Sea
Caracas
VEN.
COL.
Canary Is.
W. SAHARA
ALGERIA
LIBYA
SAHARA
EGYPT
Cairo
SAUDI
ARABIA
Mecca
OMAN
ARABIAN SEA
Karachi
Bombay
INDIA
Calcutta
B. of Bengal
Madras
SRI LANKA (CEYLON)

CAPE VERDE
Dakar
SENEGAL
MAURITANIA
MALI
NIGER
CHAD
SUDAN
ETHIOPIA
SOMALIA
G. of Aden
MALDIVES

SIERRA LEONE
LIBERIA
TOGO
UPPER VOLTA
NIGERIA
GHANA
Lagos
CAM.
CENT. AFR. REP.
UGANDA
KENYA
Nairobi
Mogadishu

Equator
Belém
Manaus
GABON
CONGO
Brazzaville
Kinshasa
ZAIRE
TANZANIA
Dar es Salaam
SEYCHELLES
Equator

SOUTH
AMERICA
BRAZIL
Natal
Salvador
Ascension (Br.)
Luanda
St. Helena (Br.)
ANGOLA
ZAMBIA
MALAWI
MOZAMBIQUE
INDIAN

agasta
PERU
BOLIVIA
LA PAZ
PAR.
Rio de Janeiro
São Paulo
Brasília
SOUTH
Tropic of Capricorn
NAMIBIA (S.-W. AFRICA)
Windhoek
BOTS.
ZIM.
Maputo
MADAGASCAR
MAURITIUS
Reunion (Fr.)

iago
Córdoba
URU.
Asunción
Montevideo
Buenos Aires
ATLANTIC
Johannesburg
SOUTH AFRICA
LESOTHO
Cape Town
C. of Good Hope
OCEAN

ARGENTINA
CHILE
Tristan da Cunha (Br.)
Gough I. (Br.)
OCEAN
Pr. Edward Is. (S. Afr.)
Crozet Is. (Fr.)
Amsterdam I. (Fr.)
St. Paul I. (Fr.)
Kerguélen (Fr.)

FALKLAND IS. (Br.)
enas
Cape Horn
S. Georgia (Br.)
SCOTIA SEA
SOUTH ORKNEY IS. (Br.)
SOUTH SANDWICH IS. (Br.)
Bouvet I. (Nor.)
McDonald Is. (Australia)
Heard I. (Australia)

Longitude West of Greenwich
Longitude East of Greenwich
Copyright by C.S. HAMMOND & Co., N.Y.

28 — EUROPE

30° H 40° J 50° K 60° L 70° M 60°

North Cape

BARENTS SEA 70° Kolguev I. Nar'yan-Mar Pechora Ob

FINLAND Murmansk

Kola Pen. White Sea Archangel Tobolsk Irtysh

Northern Dvina Syktyvkar 3

asa Onega Perm (Molotov) Sverdlovsk Tobol

Helsinki Petrozavodsk Kirov Izhevsk Chelyabinsk

Gulf of Finland Ladoga Leningrad Kazan' Ufa Magnitogorsk 50°

Tallinn EST. S.S.R. Rybinsk Res. Yaroslavl' Gor'kiy Volga Kama

Riga Pskov Kalinin Kuybyshev Res. Orenburg (Chkalov)

LATVIAN S.S.R. Moscow Kuybyshev

LITHUANIAN S.S.R. Vitebsk Smolensk Tula Penza

Vilna WHITE Minsk

Brest RUSSIAN S.S.R. Gomel Kursk Voronezh Saratov Ural

w Kiev Khar'kov Don Volgograd (Stalingrad)

Lvov UKRAINIAN S.S.R. Astrakhan'

Chernovtsy Dnepropetrovsk Krivoy Rog Donetsk (Stalino) Rostov

Cluj Kishinev MOLD. Dnepr SEA OF AZOV Krasnodar CASPIAN SEA

ROMANIA Odessa Crimea Sevastopol' Groznyy

Ploieşti Cauc. GEORGIAN S.S.R. Baku 40°

Bucharest Varna BLACK SEA Tbilisi AZERBAIDZHAN S.S.R.

Sofia Istanbul ARMENIAN S.S.R.

BULGARIA Plovdiv Bosporus Erivan

Sea of Marmara 5

Thessaloníki Dardanelles

Larísa Izmir

Athens TURKEY 30° 40°

AEGEAN SEA

Iráklion

Crete

EUROPE — 29

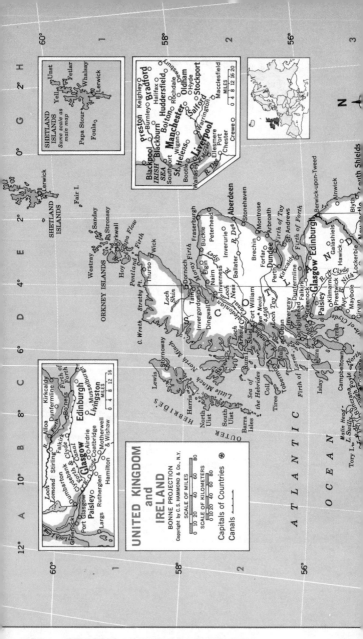

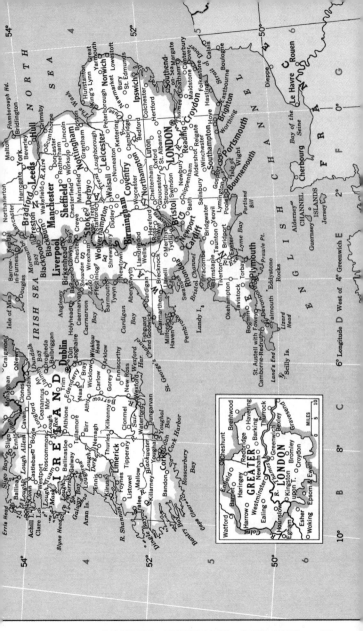

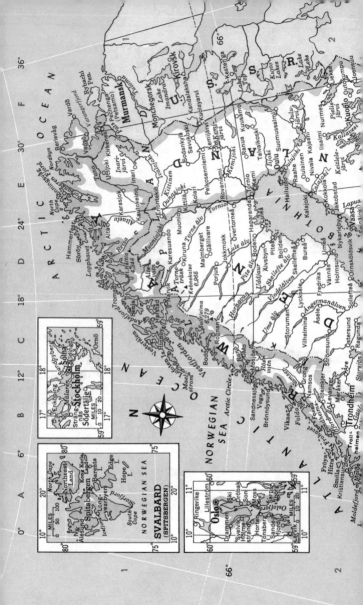

NORWAY, SWEDEN, FINLAND and DENMARK

CONIC PROJECTION

Copyright by C. S. HAMMOND & Co., N. Y.

SCALE OF MILES

KILOMETRES

Capitals of Countries ....
International Boundaries ....
Canals ....

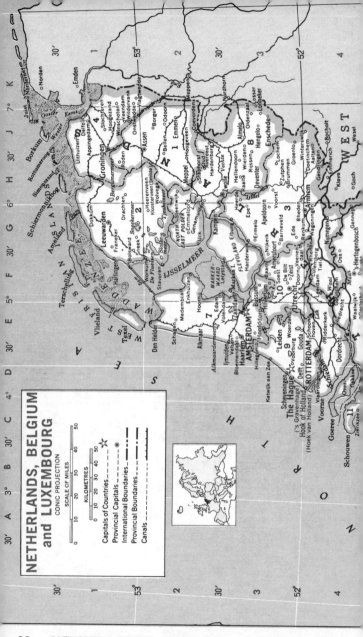

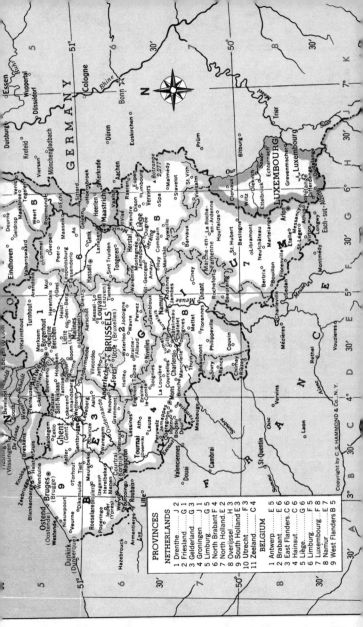

PROVINCES

NETHERLANDS

1 Drenthe	J 2
2 Friesland	G 1
3 Gelderland	G 3
4 Groningen	J 1
5 Limburg	G 5
6 North Brabant	F 4
7 North Holland	E 2
8 Overijssel	H 3
9 South Holland	F 3
10 Utrecht	F 3
11 Zeeland	C 4

BELGIUM

1 Antwerp	E 5
2 Brabant	E 6
3 East Flanders	C 6
4 Hainaut	C 6
5 Liège	G 6
6 Limburg	F 5
7 Luxembourg	F 8
8 Namur	E 7
9 West Flanders	B 5

Copyright by C. S. HAMMOND & CO., N.Y.

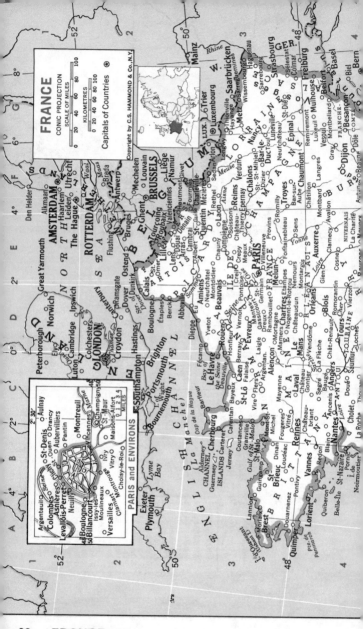

FRANCE

CONIC PROJECTION

SCALE OF MILES

0 20 40 60 80 100

KILOMETRES

0 20 40 60 80 100

⊕ Capitals of Countries

Copyright by C.S. HAMMOND & CO., N.Y.

PARIS and ENVIRONS

MILES
0 1 2 3 4

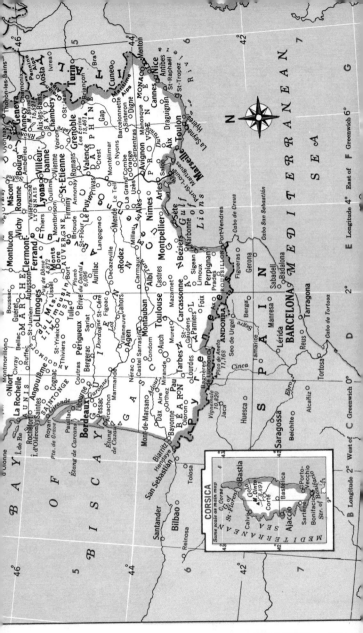

FRANCE — 39

40 — SPAIN, PORTUGAL

SPAIN, PORTUGAL — 41

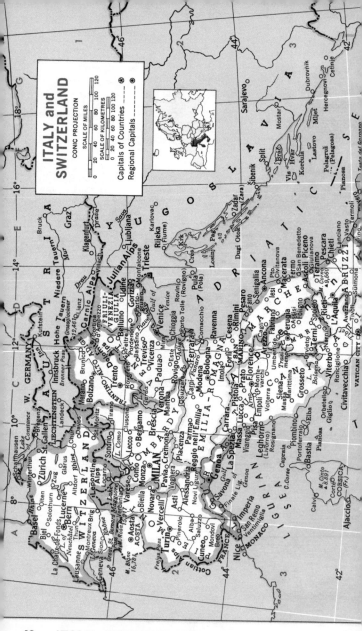

ITALY and SWITZERLAND

CONIC PROJECTION

SCALE OF MILES
0 20 40 60 80 100 120

SCALE OF KILOMETRES
0 20 40 60 80 100 120

Capitals of Countries ⊛ ⦿
Regional Capitals ●

ITALY, SWITZERLAND — 43

44 — POLAND

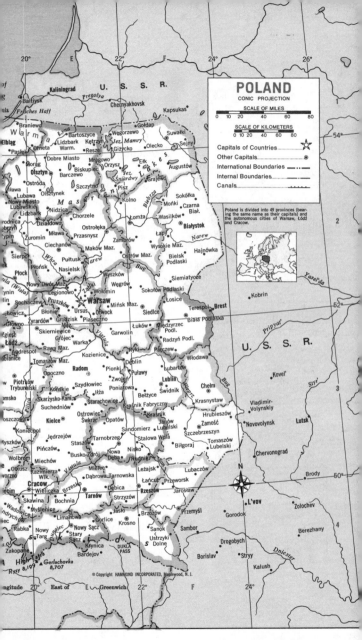

46 — AUSTRIA, CZECHOSLOVAKIA, HUNGARY

AUSTRIA, CZECHOSLOVAKIA and HUNGARY

Conic Projection

SCALE OF MILES

0 20 40 60 80 100

SCALE OF KILOMETRES

0 20 40 60 80 100

Capitals of Countries ⊛
International Boundaries — ∙∙∙ —
Canals — ∙ —

BALKAN STATES

CONIC PROJECTION

SCALE OF MILES
0   25   50   100   150   200   250

SCALE OF KILOMETRES
0   60   120   180   240   300

Capitals of Countries........ ⊛ Canals
International Boundaries.......

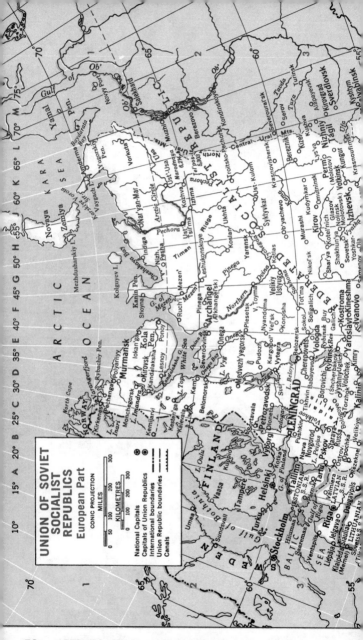

UNION OF SOVIET SOCIALIST REPUBLICS
Asiatic Part

LAMBERT AZIMUTHAL EQUAL-AREA PROJECTION

SCALE OF MILES

| 0 | 150 | 300 | 600 | 900 |

SCALE OF KILOMETERS

| 0 | 300 | 600 | 900 |

National Capitals..........................☆
International boundaries......–·–·–
Union Republic boundaries ....
Canals ....

ARCTIC

Svalbard (Spitsbergen) (Nor.)

Komsomolets

Franz Josef Ld.

Octob

Novaya Zemlya

Revolution

KARA SEA

Gol'chikha

Yamal Pen.

Dudinka

Noril'sk

Salekhard

Tazovskoye

Igarka

EUROPEAN

Voronezh

MOSCOW

RUSSIAN

UNION OF SOVIET FEDERA

Yenisey

Turukha

Pechora

Berezniki

Berezovo

Ob'

Surgut

Volga

Perm

Nizhniy Tagil

Tyumen'

Tobol'sk

Narym

Yeniseysk

Volgograd

Zlatoust

Sverdlovsk

Chelyabinsk

Troitsk

Ishim

Tara

Barabinsk

Anzhero-Sudzhensk

Kemerovo

Kra

Ural

Aktyubinsk

Tobol

Kustanay

Omsk

Novosibirsk

Tomsk

Leninsk-Kuznetskiy

Minusin

CASPIAN SEA

Gur'yev

Chelkar

KAZAKH

L. Tengiz

Irtysh

Tselinograd

Semipalatinsk

Prokop'yevsk

Novokuznetsk

Gorno-Altaysk

Krasnovodsk

Aral Sea

Kazalinsk

Leninsk

Syr-Dar'ya

Kzyl-Orda

Karaganda

S. S. R.

L. Balkhash

Barnaul

Ayaguz

Zaysan

Alakol'

Kova

Ashkhabad

Khiva

UZBEK

Amu-Dar'ya

Bukhara

Tashkent

Namangan

Kokand

Andizhan

Frunze

Panfilov

Yining

Ürümqi

Ti

TURKMEN S.S.R.

Meshed

Mary

Samarkand

TADZHIK S.S.R.

KIRGIZ

Alma-Ata

Issyk

Ti en

Shan

SINKIANG

IRAN

Birjand

Mazar-Sharif

Faizabad

Pamir

Okashi

Lop Nur

Hami

Maimana

Herat

Shache

Taklimakan Shamo

Yumer

Zahidan

Hindu Kush

Kabul

Peshawar

Srinagar

Yutian

Tsaidar

AFGHANISTAN

Helmand

Indus

Kandahar

Islamabad

Kunlun Shan

Qi

Kalat

Quetta

Lahore

TIBET

Chang H

Bela

Sukkur

Multan

Delhi

New Delhi

PAKISTAN

INDIA

LUZON PHILIPPINES · Mindanao

5

Samar

Davao

Leyte

Manila

CHINA

Mindoro

SOUTH

Palawan

Negros · Cebu

SEA Menado

CELEBES

Panay

Kota Kinabalu

SEA

Celebes

SABAH

Ujung Pandang

CHINA SEA

BRUNEI

SARAWAK

Borneo

Kuching

M A L A Y S I A

Hainan (Port.)

Hanoi · Haiphong

Gulf of Tonkin

LAOS

Mekong R.

Vientiane

THAIL'D

Bangkok

G. of Thailand

Phnom Penh

Ho Chi Minh City (Saigon)

VIETNAM

Kuala Lumpur

SINGAPORE

Balikpapan

Banjarmasin

MAL.

JAVA SEA

Palembang

Sumatra

Jakarta

Medan

Str. of Malacca

FLORES SEA

Flores

Surabaya

Bali

BANDA SEA

Ceram

Timor

Macassar Str.

Celebes

I N D O N E S I A

S U N D A

Java

Christmas I. (Austr.)

George Town

Padang

20°

AUSTRALIA

Broome

Perth

40°

Copyright by C.S. HAMMOND & Co., N.Y.

BURMA

Mandalay

Rangoon

Moulmein

Sittwe

BAY

OF

BENGAL

Andaman Is. (India)

Nicobar Is. (India)

Equator

Cocos Is. (Austr.)

Tropic of Capricorn

N D I A

Imphal

Patna

Chandernagor

Dacca

BANGL.

Calcutta

Nagpur

Vanam

Madras

Pondicherry

Karikal

Indore

SRI LANKA (CEYLON)

Kandy

Colombo

Ahmadabad

Daman

Poona

Hyderabad

Bangalore

Mahe

Madurai

Bombay

Panjim

I N D I A N

O C E A N

Diego Garcia
BRIT. IND. OC. TERR.

Male

MALDIVES

N

Laccadive Is. (India)

ARABIAN

SEA

Muscat

G. of Kxaii

Kuria Muria Is.

Socotra (P.D.R. Yemen)

YEMEN
ARAB REP.

P.D.R. YEMEN

Mukalla

Hodeida

Taizz

G. of Aden

Djibouti

Aden

AFRICA

SEYCHELLES

Madagascar

20°

ASIA
LAMBERT AZIMUTHAL
EQUAL-AREA PROJECTION
SCALE OF MILES
0   300   600   900   1200
SCALE OF KILOMETERS
0   300   600   900   1200

Capitals of Countries .... ⊛
International Boundaries ..... — ·· —
Canals ....

40°

E...

60°

F   Longitude   80°   East   of   G   Greenwich   100°   H   120°   J

56 — NEAR AND MIDDLE EAST

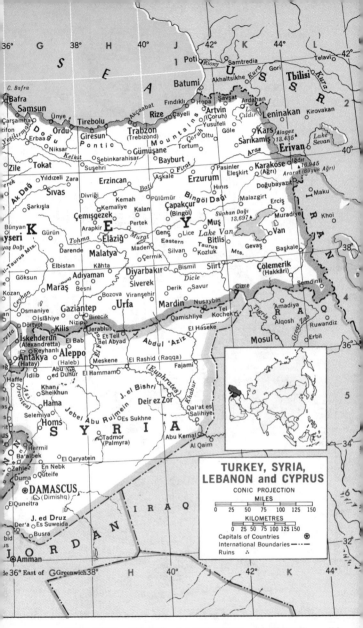

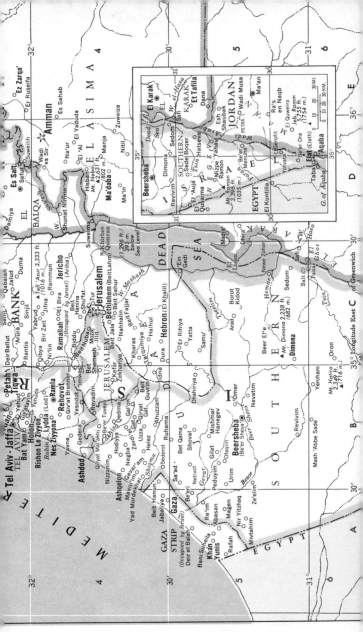

IRAN and IRAQ
CONIC PROJECTION

MILES
0 25 50 100 150 200

KILOMETRES
0 25 50 100 150 200

Capitals of Countries   ⊛
International Boundaries   — — —
Ruins   ∴

Elevations in Feet

IRAN, IRAQ — 63

64 — INDIAN AND INDOCHINESE PENINSULAS

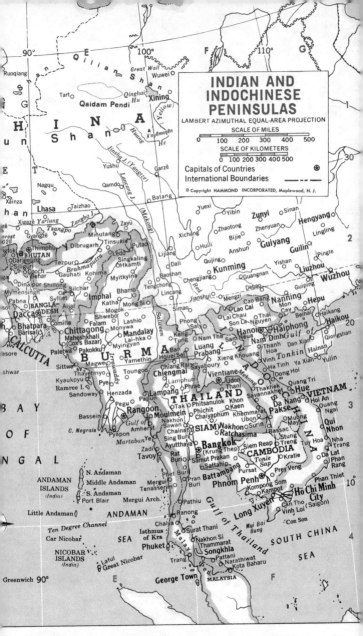

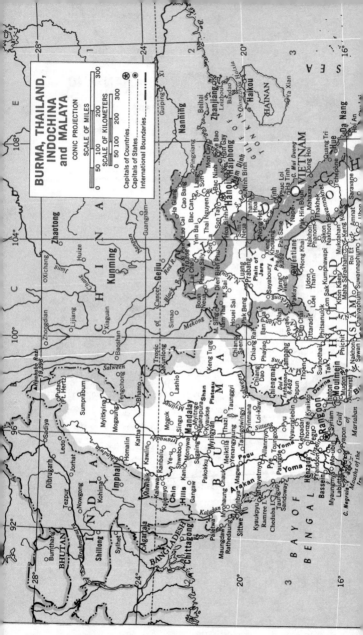

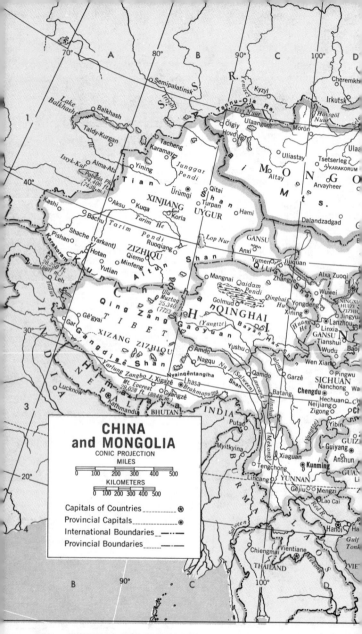

CHINA and MONGOLIA

CONIC PROJECTION

MILES
0   100   200   300   400   500

KILOMETERS
0   100   200   300   400   500

Capitals of Countries .... ⊛
Provincial Capitals .... ⊙
International Boundaries __.__.__
Provincial Boundaries _____

CHINA, MONGOLIA — 69

A 124° B 128° C 132° D 136°

1 44° Changchun

Mudanjiang

Jilin

Lake Khanka

Spassk-Dal'niy

Tetyukhe

Liaoyuan

Suifenhe

U. S. S. R.

Tunhwa

Ussuriysk

Artem

Margaritovo

Hailong

Songhua Hu

Tumen

Ol'ga

Fushun

Yanji

Hunchun

Suchan

Tonghua

Hoeryŏng

Vladivostok

Manp'o

Paektu (Baektu) 9,003

Aoji-dong

Najin

Peter the Great Bay

Ch'osan

Changbai sanmaek

Musan

Raskino

Ch'ŏngjin

Kwanmo 8,337

Sinŭiju

Kanggye

Kapsan

Nanam

Chuŭronjang

Changjin Res.

Paekam

Puksubaek 8,274

P'ungsan

Kilchu

SEA

N

OF

Hamhŭng

Hŭich'on

Hongwŏn

Kimch'aek

Tanch'ŏn

NORTH

West Korea Bay

Anju

Yŏnghŭng

East Korea Bay

KOREA

Namp'o

P'yŏngyang

Wŏnsan

JAPAN

Anak

Sŏngnim

Sariwŏn

Changjŏn

Haeju

Kaesŏng

P'yŏnggang

Inch'ŏn

P'anmunjŏm

Ch'ŏrwŏn

Yangyang

Seoul

Ch'unch'ŏn

Wŏnju

Kangnŭng

Ullŭng (Dagelet)

Samch'ŏk

Tok-to (Takeshima) (Claimed by S. Kor. & Jap.)

Ch'ungju

SOUTH

Wajim

Sŏsan

Ch'ŏngju

KOREA

Noto Pen

Taejŏn

Andong

Nana

Kunsan

Kŭm

Taebaek 5,121

Oki Is.

Dogo

Kanazawa

Chŏnju

Muju

Kimch'ŏn

Dozen

Komatsu

Mokp'o

Chiri 6,283

Sangju

P'ohang

Izumo

Tottori

Fukui

Kwangju

Namwŏn

Chinju

Taegu

Ulsan

Miryang

Hamada

Yonago

Tsuyama

Wakasa Bay

Takao

Sunch'ŏn

Masan

Chinhae

Pusan

Miyoshi

Okayama

Himeji

Kyoto

Kobe

Tsuru

Tsu Is.

Köje

Yamaguchi

Masuda

Tsuyama

Biwa

Ōsaka

Wakayama

Cheju Strait

Shimonoseki

Ube

Hiroshima

Kure

Takamatsu

Awaji

Izumisano

Shingu

Kitakyushu

Suo Sea

Iyo

Matsuyama

Ishizuchi 6,499

Kōchi

Muro

Cheju (Quelpart)

Fukuoka

Iki

Yawatahama

SHIKOKU

Pt. Shiono

Karatsu

Beppu

Uwajima

Susaki

Muroto

Goto Is.

Saga

Oita

Sukumo

Sasebo

Ōmuta

Kumamoto

Nagasaki

Nobeoka

Amakusa Is.

Hitoyoshi

KYUSHU

PACIFIC

Kobayashi

Miyazaki

EAST

Miyakonojo

CHINA

Kagoshima

Nichinan

OCEAN

SEA

Kanoya

B 128° C 132° D 136°

70 — JAPAN, KOREA

Map of Japan and Korea with detailed place names.

Main map features:

SEA OF OKHOTSK

KURILIS.
U.S.S.R.

Pt. Soya

Rebun
Rishiri
Wakkanai
Esashi

Teshio
Haboro
Nayoro

Mombetsu
C. Shiretoko
Abashiri
Kitami
Shibetsu
Bekkai
Nemuro

Iturup

Kunashir

Shikotan (Occ. by U.S.S.R.)

Habomai Is.
Pt. Nosappu

Ruinoi
Mashike

Asahikawa
Ashibetsu
Akabira

Shari

Otaru
Ishikari Bay
Yoichi
C. Kamui
Iwanai

Sapporo
Tobetsu
Ebetsu
Bibai
Yubari
Iwamisawa

Obihiro
Ikeda

Kushiro

Akkeshi

HOKKAIDO

Suttsu
Yakumo
Motsutano
Okushiri

Uchiura Bay
Muroran
Tomakomai
Urakawa

Hiroo
C. Erimo

Esashi
Mori
Hakodate
C. Esan
Pt. Esan
C. Shiriya

Matsumae
Tsugaru Str.
Mutsu
Mutsu Bay
Aomori

Misawa
Hachinohe

Hirosaki
Odate
Kosaka
Kuji

Noshiro
C. Nyudo
Oga
Hachiro Gata
Akita
Honjo

Morioka
Iwate 6,696
Miyako

Omagari
Yokote
Mizusawa
Kamaishi

Chokai 7,316
Sakata
Tsuruoka

Shinjo
Furukawa
Kesennuma

Yamagata
Yuzawa
Ishinomaki
Shiogama
Onagawa

Niigata
Yonezawa
Fukushima
Sendai

Sanjo
Nagaoka
Koriyama
Shirakawa
Aizuwakamatsu

Shirane 8,458
Nikko
Hitachi
Iwaki

HONSHU

Takasaki
Kiryu
Maebashi
Utsunomiya
Mito
Tsuchiura

Kofu
Fuji 12,389
Kumagaya
Tone

TOKYO
Chiba
Choshi
C. Inubo

Yokohama
Katsuura

Yokosuka
Tateyama
Nojima

Numazu
Ito
Atami
Sagami Sea
O-Shima

Shimoda

Suruga Bay
Shizuoka

PACIFIC OCEAN

Kyoto/Osaka inset (lower left):

Maizuru
Takatsuki
Toyonaka
Suita
Hirakata
Uji
Otsu

Kyoto

Osaka
Nara
Yao
Tenri
Yamatokoriyama
Izumiotsu
Sakai

135°30'
34°30'

Tokyo inset (upper right):

MILES
0 5 10

Mitsukaido

Omiya
Noda
Kashiwa
Hanno
Warabi
Kawaguchi
Urawa
Ichikawa
Tokorosawa

Funabashi
Hachioji
Chiba
Machida
Tokyo Bay
TOKYO

Hiratsuka
Kawasaki

Yokohama
Kamakura
Kisarazu

Yokosuka
Amaha
Sagami Bay
Miura

36°
35°
30'
139°30'
140°

NAMPO SHOTO inset (center):

NAMPO SHOTO

142° Same scale as main map

BONIN ISLANDS
(OGASAWARA GUNTO)

Muko
Nishino
Chichi
Omura
Haha

VOLCANO ISLANDS
(KAZAN-RETTO)

Kita Iwo

Iwo

Minami

PACIFIC OCEAN

28°
26°
24°
28°

Ryukyu / Kyushu area (right):

KYUSHU

EAST CHINA SEA

Tanega
Yaku
Osumi Is.
Kuchino
Suwanose
Tokara Is.

RYUKYU NANSEI SHOTO

Naze
Kikai
Amami-O-Shima
Tokuno
Okino-Erabu
Yoron

Ie
Motobu
Nago
Kume
Naha
Okinawa
Kerama Is.
Shuri
Itoman
Okinawa Is.

PACIFIC OCEAN

40°
28°
36°
128°
124°

Sakishima inset (lower center):

Yaeyama Is.
Miyako Is.
Miyako
Hirara
Tarama
Shimoji
Yonaguni
Iriomote
Ishigaki

SAKISHIMA
Tropic of Cancer

ISLANDS

PACIFIC

Same scale as main map

124°
24°

Legend box:

JAPAN and KOREA

CONIC PROJECTION

Copyright by C. S. HAMMOND & CO., N.Y.

SCALE OF MII

0   50   100   150

KILOMETRES

0   50   100   200   300

Capitals of Countries ....⊛
International Boundaries ....

Grid labels:

140° 144° 148° 152°
1
44°
40°
3
28°
36°
4
32°
5

East of   140°   Greenwich   144°   G   148°   H

F   G   H

JAPAN, KOREA — 71

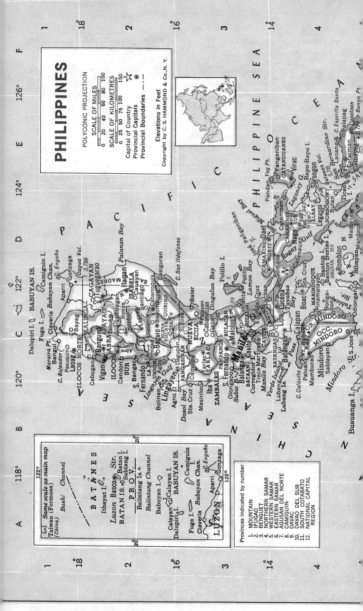

PHILIPPINES

POLYCONIC PROJECTION

SCALE OF MILES
0 20 40 60 80 100

SCALE OF KILOMETRES
0 25 50 75 100 150

Capital of Country ⭐
Provincial Capitals ◉
Provincial Boundaries --- ---

Elevations in Feet
Copyright by C. S. HAMMOND & Co., N. Y.

Provinces indicated by number
1. MOUNTAIN
2. IFUGAO
3. BENGUET
4. NORTHERN SAMAR
5. WESTERN SAMAR
6. EASTERN SAMAR
7. AGUSAN DEL NORTE
8. CAMIGUIN
9. DAVAO
10. DAVAO DEL SUR
11. SOUTH COTABATO
12. NATIONAL CAPITAL REGION

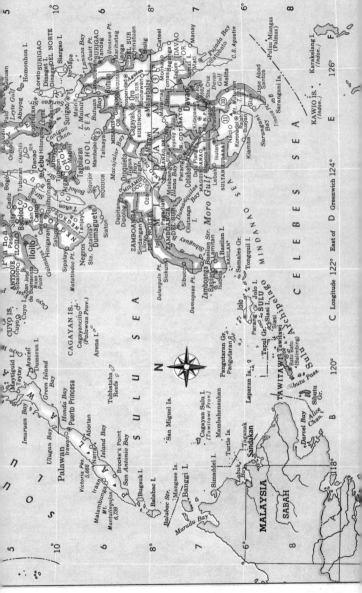

74 — SOUTHEAST ASIA

SOUTHEAST ASIA
LAMBERT AZIMUTHAL EQUAL-AREA PROJECTION

SCALE OF MILES
0   100   200   400   600

SCALE OF KILOMETRES
0   100   200   400   600

Capitals of Countries----------------⊛
Administrative Center----------------◉
International Boundaries----------------
Territorial Boundaries----------------

JAVA inset

Jakarta · Serang · Bogor · Bandung · Sukabumi · Indramaju · Cirebon · Tegal · Pekalongan · Semarang · Kudus · Rembang · Surakarta · Solo · Madiun · Blitar · Malang · Surabaya · Madura · Pamekasan · Probolinggo · Banyuwangi · Pasuruan · Kediri · Yogyakarta · Magelang · Ciamis · Cilacap · Mt. Slamet 11,247 · Mt. Semeru 12,060 · Karimunjawa Is. · Bawean · JAVA SEA · INDIAN OCEAN · Madura Str.

MILES
0 25 50

Main map labels

Taiwan (Formosa) (China) · Batan Is. · Babuyan Is. · Laoag · Vigan · Tuguegarao · Baguio · Cabanatuan · Cagayen · Tarlac · LUZON · Manila · Batangas · Mindoro · Catanduanes · Legaspi · Samar · Masbate · Catbalogan · Tacloban · Leyte · Panay · Iloilo · Bacolod · Cebu · Bohol · Negros · PHILIPPINES · Puerto Princesa · Palamian Group · SULU SEA · Zamboanga · Basilan · Sulu Arch. · Jolo · Sandakan · Cagayan de Oro · Oroquieta · Mindanao · Davao · Moro Gulf · Davao Gulf · Sarangani · Kawio Is. · Talaud Is. · CELEBES SEA · C. Arus · Manado · Gorontalo · Sangihe Is. · Morotai · Jailolo · Ternate · Halmahera · Batjan Is. · Weda · Raja · Ampat Gr. · Obi Is. · Misool · MOLUCCA SEA · Sula Is. · Sulawesi · Banggai Arch. · Gulf of Tomini · Poso · Gulf of Tolo · CERAM SEA · Buru · Ceram · Amboina · Banda Is. · BANDA SEA · Palopo · Rantekombola 11,335 · Kendari · Butung · Bonthain · Baubau · Tukangbesi · Penyu Is. · CELEBES · Gulf of Bone · SULAWESI · FLORES SEA · Ruteng · Flores · Ende · Lombien · Alor · Dili · Wetar · Damar Is. · Babar Is. · Saumlaki · Tanimbar Is. · Aru Is. · Kai · Ewab · Waingapu · Savu Sea · Sawu Is. · Roti · Kupang · Timor · TIMOR SEA · ARAFURA SEA · Melville I. · AUSTRALIA · Wessel Is.

Sonsorol Is. · Merir I. · Tobi I. · TERR. OF THE PACIFIC ISLANDS (U. S. Trust) · Mapia Is. · Asia Is. · Waigeo Is. · Salawati · Sorong · Manokwari · Biak · Schouten Islands · C. Perkam · Mamberamo · Doberai Pen. · Fakfak · Kaimana · Sarera Bay · Maoke Mts. · Puncak Jaya 16,503 · IRIAN JAYA · Jayapura (Hollandia) · Iden Mts. · Digul · Dolak I. (Fredrik Hendrik I.) · C. Vals · Merauke · PACIFIC OCEAN

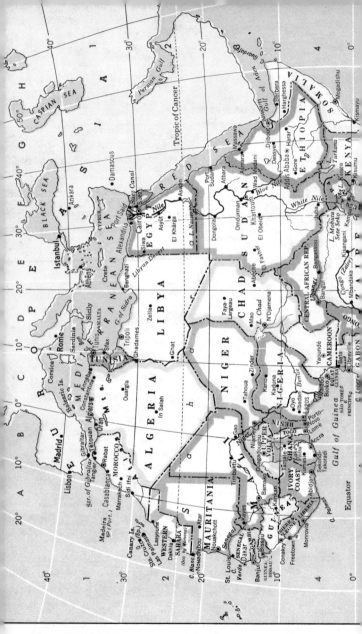

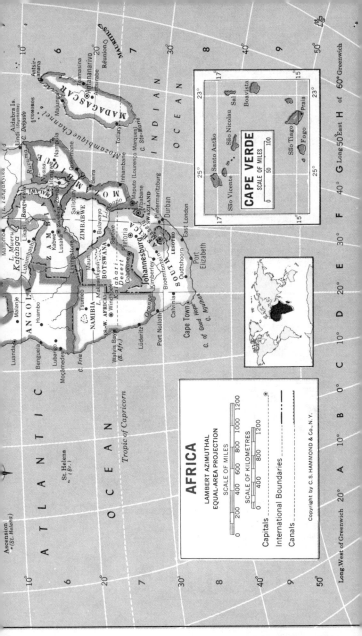

AFRICA

LAMBERT AZIMUTHAL
EQUAL-AREA PROJECTION

SCALE OF MILES

0   200   400   600   800 1000 1200

SCALE OF KILOMETRES

0   400   800   1200

Capitals .... ⊙

International Boundaries ....

Canals ....

Copyright by C. S. HAMMOND & Co., N.Y.

CAPE VERDE

SCALE OF MILES

0   50   100

Santo Antão
São Vicente
São Nicolau
Sal
Boavista
São Tiago
Praia
Fogo

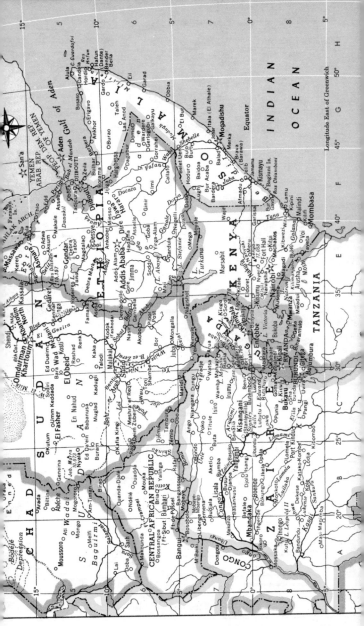

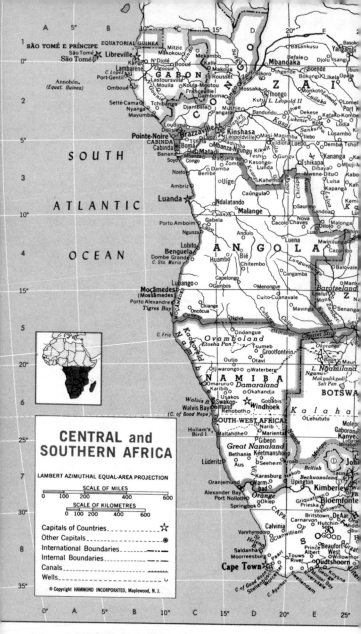

CENTRAL AND SOUTHERN AFRICA — 83

84—PACIFIC OCEAN

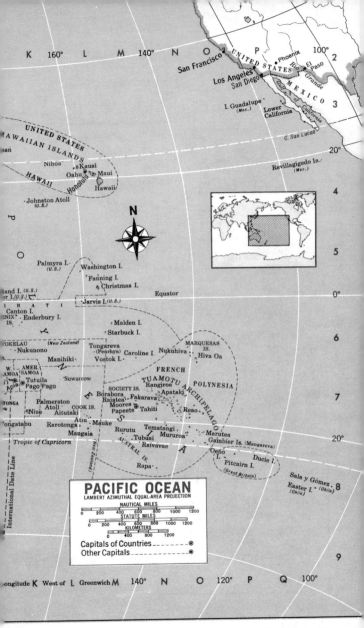

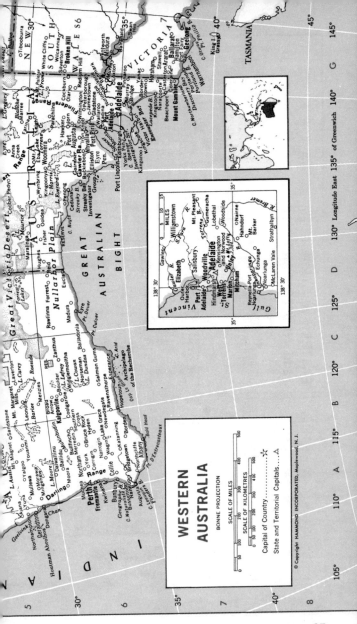

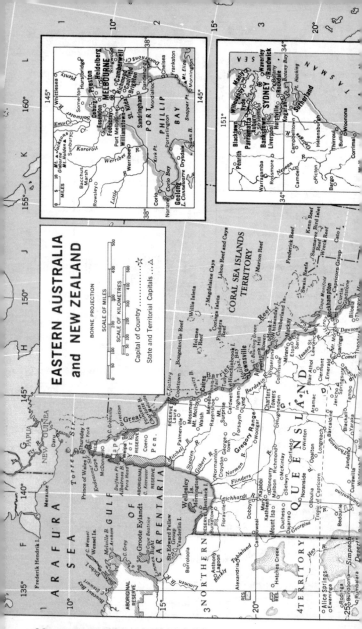

EASTERN AUSTRALIA and NEW ZEALAND

BONNE PROJECTION

SCALE OF MILES

SCALE OF KILOMETRES

Capital of Country .... ☆
State and Territorial Capitals ... △

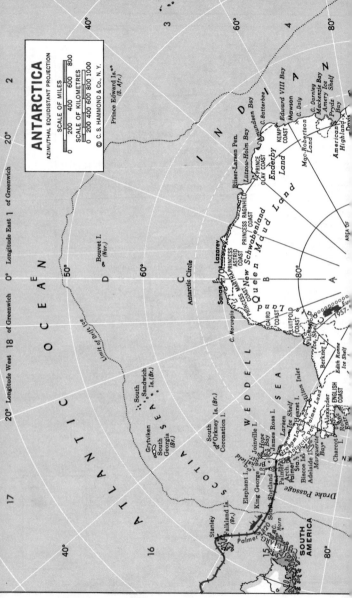

ANTARCTICA — 91

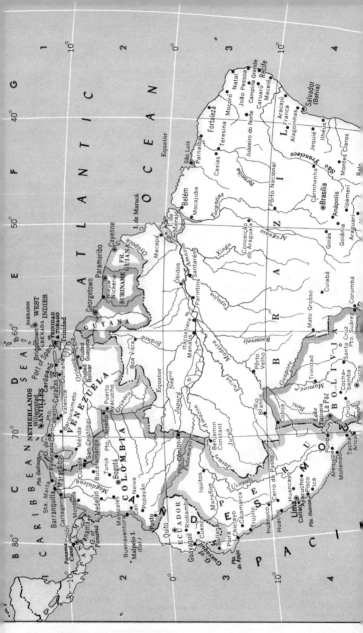

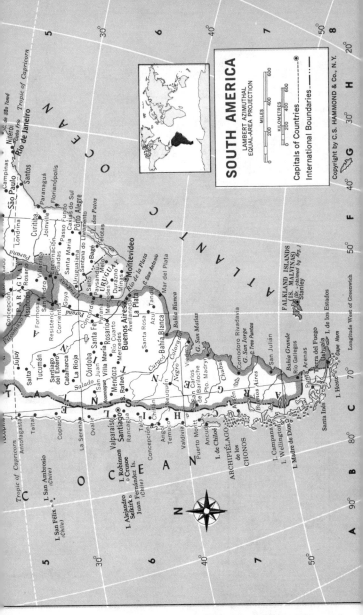

SOUTH AMERICA

LAMBERT AZIMUTHAL
EQUAL-AREA PROJECTION

MILES
0   200   400   600

KILOMETRES
0   200   400   600

Capitals of Countries .... ◉
International Boundaries ——..——..——

Copyright by C.S. HAMMOND & Co., N.Y.

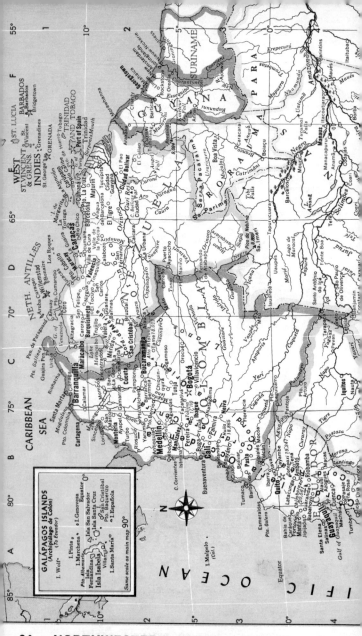

NORTHWESTERN SOUTH AMERICA

LAMBERT AZIMUTHAL EQUAL-AREA PROJECTION

SCALE OF MILES
100   200   300   400   500

SCALE OF KILOMETRES
100   200   300   400   500

Capitals of Countries .... ☆
Other Capitals .... △
International Boundaries .... — · — · —
Other Boundaries .... — — — —

Copyright by C. S. HAMMOND & CO., N.Y.

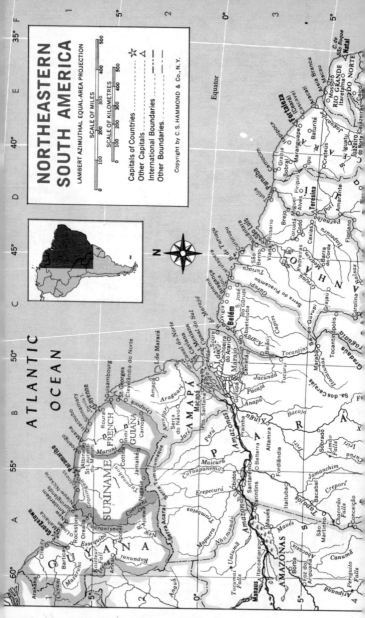

NORTHEASTERN SOUTH AMERICA

LAMBERT AZIMUTHAL EQUAL-AREA PROJECTION

SCALE OF MILES
100   200   300   400   500

SCALE OF KILOMETRES
100   200   300   400   500

Capitals of Countries .... ☆
Other Capitals .... △
International Boundaries .... —·—·—
Other Boundaries .... ——

Copyright by C. S. HAMMOND & CO., N.Y.

N

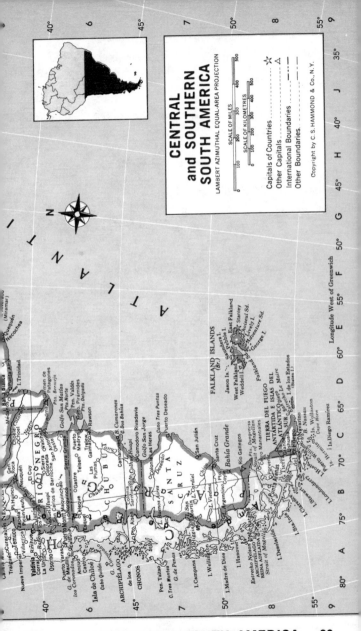

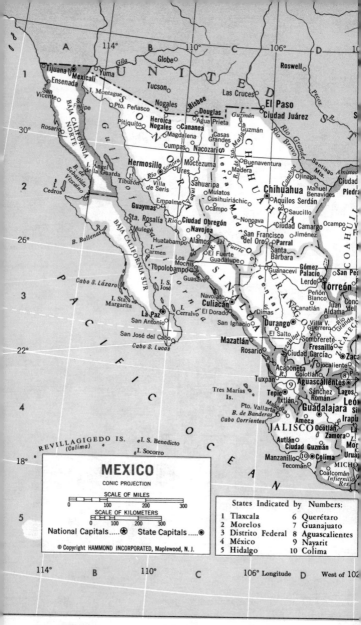

Inset map (top)

HIDALGO
Chignahuapan
Misantla
Teziutlán
Altotonga
GULF OF
Coyotepec
Zumpango
Teotihuacán
Apan
San Salvador el Seco
Cuyoaco
Villa Ursulo Galván
Azcapotzalco
Calpulálpam
Libres
Nauhcampatepetl 14,045
Jalapa
MEXICO
Texcoco
TLAXCALA
Jico
Rio Pescados
Coatepec
MEXICO CITY
S. Martín Texmelucan
Apizaco
R. J. García
Veracruz
Coyoacán
Xochimilco
D.F.
Amecameca
Tlaxcala
Villa Vicente Guerrero
Huatusco de Chicuellar
Soledad de Doblado
Cholula
Puebla
Citlaltépetl 18,855
Córdoba
Cuernavaca
Popocatépetl 17,887
Tepeaca
Ciudad Serdán
Orizaba
MORELOS
Atlixco
Tecamachalco
Catzingo
Ciudad Mendoza
Zongolica
Rio Blanco
Cuautla
Jonacatepec
Izúcar de Matamoros
Morelos
Xochitepec
San Gabriel
Chapulco
Jojutla

MILES
0 10 20 30

Main map

Corpus Christi
Laredo
Falcon Res.
El Azúcar Res.
Brownsville
Reinosa
Matamoros
Cadereyta
Montemorelos
Laguna
Linares
Madre
Hidalgo
Santander Jiménez

GULF OF MEXICO

Tropic of Cancer

Ciudad Victoria
Tula
Aldama
C. Mante
Ciudad Madero
Tampico
I. Pérez
Cayo Arenas
Cabo Catoche
Campeche Bank
POTOSI
Cárdenas
Pánuco
Progreso
Hunucmá
Tizimín
YUCATAN
Jalpan
Tantoyuca
Mérida
Chichén-Itzá
Valladolid
Cozumel
Huejutla
Mineral del Monte
Tuxpan
Papantla de Olarte
Bay
Cayo Arcas
of Campeche
Halachó
Calkiní
Ticul
Uxmal (Ruin)
Tekax
Hopelchén
Felipe Carrillo
L. Bacalar
Pachuca
Tulancingo
Jalapa
Champotón
Sabancuy
Chetumal
MEXICO CITY
Teziutlán
Tlaxcala
VERACRUZ
Veracruz
Orizaba
San Andrés Tuxtla
Coatzacoalcos (Pto. México)
Carmen
Lag. de Términos
CAMPECHE
Xcalak
Puebla
Alvarado
Frontera
Palenque
Belmopan
Belize City
Tehuacán
Tierra Blanca
Tuxtepec
Minatitlán
Villahermosa
TABASCO
Teapa
Tenosique
Stann Creek Town
San Gabriel
Huajuapan
Matías Romero
Ciudad de las Casas (S. Cristóbal)
Flores
Gulf of Honduras
GUERRERO
Tixtla
Mitla (Ruin)
Tuxtla Gutiérrez
CHIAPAS
Cobán
Tlaxiaco
Oaxaca
OAXACA
Comitán
Iguala
Tehuantepec
Juchitán
Salina Cruz
Tonalá
Huehuetenango
Zacapa
HONDURAS
Ometepec
de
Sur
Huixtla
Quezaltenango
Guatemala
Sta. Rosa
Sta. Maldonado
Puerto Ángel
Gulf of Tehuantepec
Tapachula
Madre
GUATEMALA
BELIZE
QUINTANA ROO

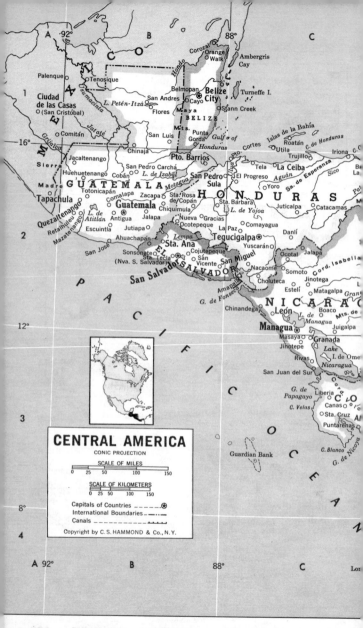

JAMAICA
Kingston

Pedro
Bank

Pedro Cays
(Jam.)

Morant Cays
(Jam.)

Rosalind
Bank

Banco de
Serranilla
(Col.)

Bajo Nuevo
(Col.)

Gorda
Bank

Laguna de
Caratasca
Caratasca

Cabo Gracias a Dios

Cayos Miskitos

Serrana Bank
(Col.)

Pto. Cabezas
(Bragman's Bluff)

Quita Sueno Bank
(Col.)

Roncador Cay
(Col.)

Prinzapolka

I. de Providencia
(Col.)

Laguna de
Perlas

I. de
San Andrés
(Col.)

Corn Is.
(Nic.)

Cayos de
Albuquerque
(Col.)

N

Pta. del Mono

San Juan del Norte
(Greytown)

San José
Cartago

Limon

Pta. Manzanillo

G. de San Blas

Colon

Bocas del Toro

Mosquito Gulf

Panama
Canal

Panamá

Serranía del Darién

G. de Urabá

Lag. de
Chiriquí

Chorrera

Ta amanca

Pto. Cortes

P A **N** A **M** A

Golfito

Serr. de
Tabasará

Penonomé

David

La Palma

G. Dulce

Aguadulce

Gulf
of

Arch. de
las Perlas

Turbo

Pto. Armuelles

G. de
Chiriquí

Santiago

G. de Parita

El Real

Chitré

Panamá

Pta. Burica

Pen. de
Azuero

Las Tablas

COLOMBIA

I. Coiba

West of Greenwich

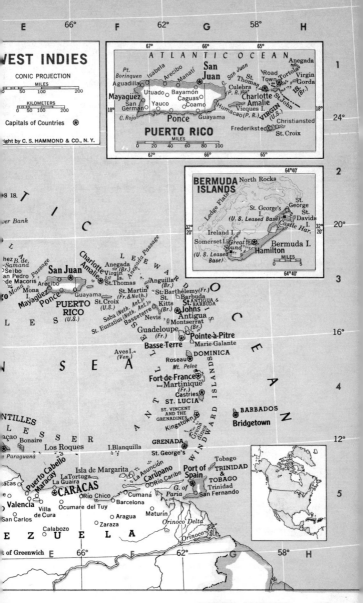

BEAUFORT SEA

UNITED STATES

ALASKA

YUKON

Yukon
Fairbanks
Circle
Fort Yukon
Tanana
Porcupine
Arctic Creek
Aklavik
Inuvik
Arctic Red
River
Tuktoyaktuk
Paulatuk
Coppermine
Holman I.
Read I.
Banks Island
Amundsen Gulf
Mackenzie Bay
Peel
Dawson
Mayo
Ross
Selkirk
Pelly
Whitehorse
Stewart
Skagway
Atlin
Teslin
Watson Lake
Mt. St. Elias 18,008
Mt. Logan 19,850
Mt. Fairweather 15,300

PARRY
Melville I.
Vis. Melville Sd.
Bathurst I.
Prince of Wales I.
M'Clintock Channel
Victoria I.
DISTRICT
Cambridge Bay
Dease Str.
Bathurst Inlet
Arctic Circle
Coronation Garry L.
Baker Lake
King William
Boo

DISTRICT OF MACKENZIE
NORTHWEST
Norman Wells
Ft. Norman
Great Bear Lake
Wrigley
Ft. Simpson
Ft. Liard
Fort Nelson
Trout L.
Hay River
Ft. Providence
Great Slave Lake
Rae
Yellowknife
Ft. Reliance
Ft. Resolution
Ft. Smith
Fort Radium
Lac la Martre
Dubawnt L.
Kasba L.
Nueltin L.
Chesterfield
Padlei

ROCKY
BRITISH COLUMBIA
Prince Rupert
Prince of Wales I.
Queen Charlotte Is.
Hecate Str.
Queen Charlotte Sd.
Sitka
Wrangell
Stewart
Ketchikan
Terrace
Smithers
Burns Lake
Prince George
Vanderhoof
Quesnel
Mt. Waddington 13,260
Courtenay
Nanaimo
Vancouver Island
Victoria
Vancouver
New Westminster
Trail
Nelson
Cranbrook
Kamloops
Kelowna
Vernon
Revelstoke
Jasper
Lake Louise
Banff
Seattle
Portland
Columbia
WASH.
IDAHO
Spokane
MONT.

Churchill Ph. 12,900
St. John
Dawson Cr.
Grande Prairie
Peace River
L. Athabasca
Uranium City
Chipewyan
Wollaston L.
Cree L.
Ft. McMurray
Reindeer L.
Lynn Lake
Brochet
Amery
Nelson
Churchill
Sherridon
Flin Flon
The Pas
Cedar L.
MANITOBA
Berens River

ALBERTA
Edmonton
Vermilion
Athabasca
Meadow Lake
St. Paul
Wetaskiwin
Camrose
Lloydminster
North Battleford
Red Deer
Drumheller
Calgary
Biggar
Saskatoon
Humboldt
Melfort
Prince Albert
SASKATCHEWAN
Medicine Hat
Lethbridge
Cardston
Swift Current
Moose Jaw
Regina
Weyburn
Estevan
Souris
Yorkton
Melville
Dauphin
Kamsack
Winnipeg
Lake of the Woods
Morden
Boniface
Brandon
Manitoba
Red
Rainy River
Fort Frances
Shaunavon

N. DAK.
Bismarck
Fargo
Duluth
MINN.
S. DAK.
Pierre
Minneapolis
UNITED
Niobrara
NEBR.
IOWA
Missouri

Lake Huron
Walkerton
Newmarket
Cobourg
Orangeville
Fergus
Goderich
Listowel
Guelph
Oshawa
Toronto
MICH.
Port Huron
Stratford
Kitchener
Cambridge
Brampton
St. Catharines
Hamilton
Ontario
Niagara Falls
Woodstock
Brantford
Simcoe
Welland
Buffalo
N.Y.
Sarnia
London
St. Thomas
Welland Canal
Walkerville
Wallaceburg
Detroit
Windsor
Chatham
Sandwich
Leamington
Pt. Pelee
Lake Erie
Long Pt.
Erie

Miles
0   20   40

CANADA

CONIC PROJECTION

SCALE OF MILES

0 100 200 300 400 500

SCALE OF KILOMETRES

0 100 200 300 400 500

Capitals of Countries _____ ⊛

Provincial & Territorial
Capitals _____ ◉

Canals _____

Copyright by C.S. Hammond & Co., N.Y.

110 — UNITED STATES

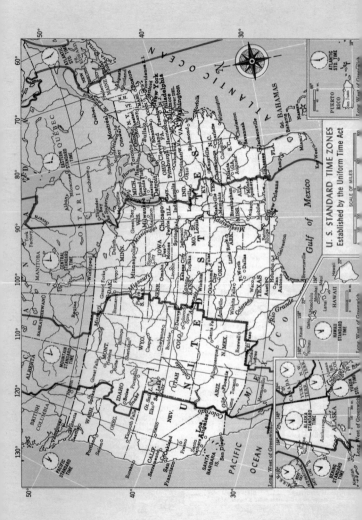

U. S. STANDARD TIME ZONES
Established by the Uniform Time Act

FACTS ABOUT THE STATES AND PROVINCES

U.S. States	Admitted to Union	Date Settled	State Nickname	State Flower
Alabama	1819	1702	Yellowhammer	Camellia
Alaska	1959	1784	The Great Land	Forget-me-not
Arizona	1912	1580	Grand Canyon	Saguaro Cactus
Arkansas	1836	1686	Land of Opportunity	Apple Blossom
California	1850	1769	Golden	Golden Poppy
Colorado	1876	1858	Centennial	Rocky Mtn. Columbine
Connecticut	1788	1633	Constitution	Mountain Laurel
Delaware	1787	1638	Diamond	Peach Blossom
Florida	1845	1565	Sunshine	Orange Blossom
Georgia	1788	1733	Peach	Cherokee Rose
Hawaii	1959	————	Aloha	Red Hibiscus
Idaho	1890	1842	Gem	Syringa
Illinois	1818	1699	Prairie	Native Violet
Indiana	1816	1732	Hoosier	Peony
Iowa	1846	1788	Hawkeye	Wild Rose
Kansas	1861	1827	Sunflower	Sunflower
Kentucky	1792	1774	Bluegrass	Goldenrod
Louisiana	1812	1699	Pelican	Magnolia
Maine	1820	1624	Pine Tree	Pine Cone & Tassel
Maryland	1788	1634	Old Line	Blackeyed Susan
Massachusetts	1788	1620	Bay	Mayflower
Michigan	1837	1668	Wolverine	Apple Blossom
Minnesota	1858	1805	North Star	Lady-slipper
Mississippi	1817	1699	Magnolia	Magnolia
Missouri	1821	1735	Show Me	Hawthorn
Montana	1889	1809	Treasure	Bitterroot
Nebraska	1867	1847	Cornhusker	Goldenrod
Nevada	1864	1851	Silver	Sagebrush
New Hampshire	1788	1623	Granite	Purple Lilac
New Jersey	1787	1617	Garden	Purple Violet
New Mexico	1912	1595	Land of Enchantment	Yucca
New York	1788	1614	Empire	Rose
North Carolina	1789	1650	Tarheel	Dogwood
North Dakota	1889	1780	Flickertail	Wild Prairie Rose
Ohio	1803	1788	Buckeye	Scarlet Carnation
Oklahoma	1907	1889	Sooner	Mistletoe
Oregon	1859	1811	Beaver	Oregon Grape
Pennsylvania	1787	1643	Keystone	Mountain Laurel
Rhode Island	1790	1636	Little Rhody	Violet
South Carolina	1788	1670	Palmetto	Yellow Jessamine
South Dakota	1889	1817	Coyote	Pasqueflower
Tennessee	1796	1757	Volunteer	Iris
Texas	1845	1682	Lone Star	Bluebonnet
Utah	1896	1847	Beehive	Sego Lily
Vermont	1791	1724	Green Mountain	Red Clover
Virginia	1788	1607	Old Dominion	American Dogwood
Washington	1889	1811	Evergreen	Western Rhododendron
West Virginia	1863	1727	Mountain	Big Rhododendron
Wisconsin	1848	1701	Badger	Wood Violet
Wyoming	1890	1834	Equality	Indian Paintbrush

Canadian Provinces	Date of Admission	Date Settled	Provincial Flower
Alberta	1905	1795	Wild Rose
British Columbia	1871	1843	Dogwood
Manitoba	1870	1812	Prairie Crocus
New Brunswick	1867	1611	Purple Violet
Newfoundland	1949	1610	Pitcher Plant
Nova Scotia	1867	1605	Trailing Arbutus
Ontario	1867	1671	White Trillium
Prince Edward Island	1873	1713	Lady's Slipper
Quebec	1867	1608	Madonna Lily
Saskatchewan	1905	1774	Wild Wood Lily

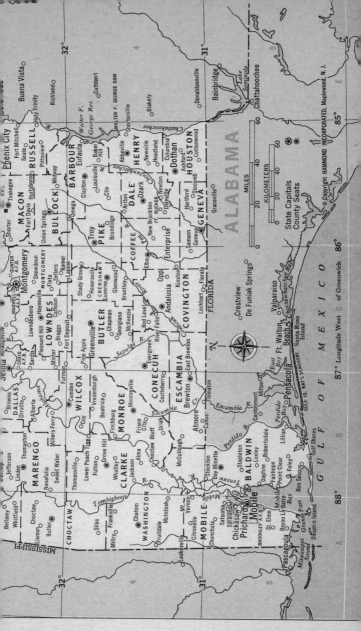

ALASKA

MILES
0   100   200   300

KILOMETERS
0   100   200   300

65° State and Provincial Capitals ⊛
Court Houses ⊙

ARCTIC

Barrow · Pt. Barrow
Wainwright
Point Lay

Chukchi
Sea

DE LONG MTS.

Kolyuchin Bay

Point Hope

Kivalina
Noatak

Unurmino

CAPE KRUSENSTERN
NAT'L MON.

BAIRD MTS.

Shishmaref
Kotzebue
Kotzebue Sound

Kiana

NOATAK NAT'L PRES.

KOBUK VALLEY NAT'L PARK

Shungnak

GATES
NAT'L
& PRES.

U.S.S.R.

Uelen
Lavrentiya

Diomede Is.
Bering Is.
C. Pr. of Wales

Noorvik

Arctic

Koyuk

Huslia

BERING LAND BRIDGE
NAT'L PRES.

Selawik

Providenya

Seward
Pen.

Teller
White
Mountain

Nome

Koyuk

Nulato

Galena
Ruby

Gambell
Savoonga

St. Lawrence I.

Norton
Sound

Shaktoolik

Kaltag

McGrath

Northeast C.

Unalakleet

CEN

Southeast C.
Emmonak
Alakanuk

Stuart I.

Shageluk

Holy Cross

KUSKOKWIM MTS.

Hall I.

Scammon Bay

Mtn.
Village

Anvik

St. Matthew I.

Hooper Bay

Chevak

Russian Mission

Aniak

Sleetmute

BERING

Tununak
Mekoryuk

Bethel

Kwethluk

Nunivak I.

Kipnuk

Eek

Quinhagak

LAKE CLARK NAT'L
PARK & PRES.

Nondalton

Kwigillingok

Kuskokwim Bay

Goodnews
Bay

C. Newenham

Aleknagik

Newhalen

Iliamna

Togiak

Dillingham

Hagemeister
I.

KATMAI
NAT'L PARK & PRES.

St. Paul I.

PRIBILOF IS.

Egegik

Naknek

Shelikof

SEA

55°

St. George I.

Bristol   Bay

Pilot Pt.

ANIAKCHAK
N.M. & PRES.

Port Moller

Alaska

Chignik

Karluk

Trinity Is.

Perryville

Chirikof I.

ISLANDS

Unimak Pass

Unimak I.

Sand Point

King Cove

Shumagin
Is.

SOUTH

Akutan

Dutch Harbor

Unalaska I.

Umnak Pass

Sanak I.

ALEUTIAN

Unalaska

FOX   IS.

Umnak
I.

Is. of the
Four Mountains

Nikolski

Seguam

PACIFIC   OCEAN

Amlia I.

© Copyright HAMMOND INCORPORATED, Maplewood, N. J.

170°   160°

116 — ALASKA

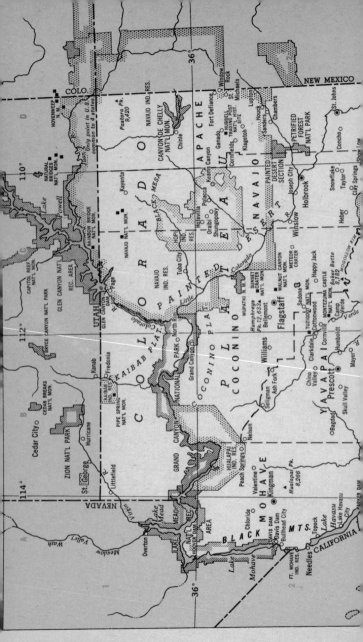

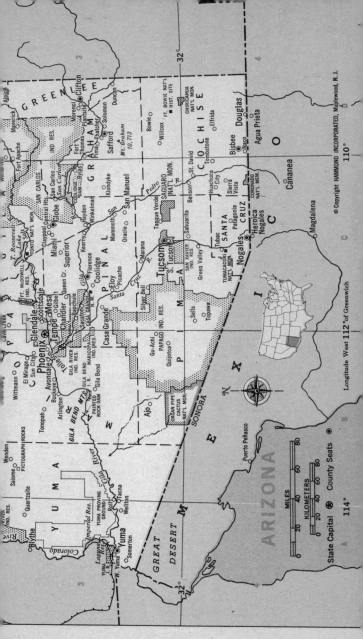

GREENLEE
GRAHAM
COCHISE
PINAL
PIMA
YUMA
PAPAGO IND. RES.
MARICOPA
GILA
SANTA CRUZ

Phoenix
Tucson
Yuma
Nogales
Douglas
Bisbee
Tombstone
Safford
Clifton
Globe
Florence
Casa Grande
Ajo
Sells
Willcox
Benson
Patagonia

SAGUARO NAT'L MON.
TONTO NAT'L MON.
CASA GRANDE RUINS NAT'L MON.
ORGAN PIPE CACTUS NAT'L MON.
CHIRICAHUA NAT'L MON.
CORONADO NAT'L MEM.
FT. BOWIE NAT'L HIST. SITE
TUMACACORI NAT'L MON.
SAN XAVIER IND. RES.
SAN CARLOS IND. RES.
GILA RIVER IND. RES.
FT. McDOWELL IND. RES.
PAINTED ROCK DAM
PICTOGRAPH ROCKS
GREAT DESERT

Mt. Graham
10,713

SONORA
MEXICO

ARIZONA

Longitude West 112° of Greenwich

MILES
0   20   40   60   80
KILOMETERS
0   20   40   60   80

State Capital ⊛   County Seats ⊙

ARIZONA — 119

ARKANSAS — 121

State Capitals ⊛ County Seats ◉

CALIFORNIA — 123

COLORADO

MILES
0   20   40   60
KILOMETERS
0   20   40   60

★ State Capitals
◉ County Seats

COLORADO — 125

Longmeadow

Webster ○ •Webster L.

30'   D   72°   E

42°

amond
kes   ○Thompsonville
ffield ●   Hazardville○ •Somersville   •Staffordville   N. Grosvenor
        Enfield   Dale
sor Locks○ ○Warehouse   Stafford   South   Thompson○ Quaddick
          ○ Point   Broad   Springs   Woodstock   Res.
ndsor ●   Brook ○   ○Putnam
omfield   Ellington○   •Willington   ○Eastford Pomfret○
Wilson○   Rockville○ ○Tolland ○Willington   Abington○   •E. Killingly
ford● ○S. Windsor TOLLAND   WINDHAM   Dayville○ •Danielson
        ○Vernon○•Talcottville ○Storrs   ○Hampton   E. Brooklyn○
RD ○ ●Manchester   Mansfield Ctr.○ Brooklyn○
        •Coventry○   River○
D   ○Andover   Willimantic○   ○Central Village
ood   ○Windham Canterbury○ ○Moosup
ewington ○Glastonbury ○Columbia   ○Plainfield ○Sterling
ocky Hill○ ○S. Glastonbury ○Hebron   Shetucket
        ○Marlborough   ○Lebanon   ○Baltic   ○Jewett City
omwell   ○Versailles
        ○Portland   ○E. Hampton ○Colchester ○Yantic   Pachaug ○Voluntown
Middletown   Pd. ○Preston City
Middlefield ○Moodus   Gardner   ○Norwich
              L.
        Durham Haddam MIDDLESEX NEW LONDON   30'
        ○ E. Haddam   ○Montville
              ○Chesterfield Uncasville○   N. Stonington○
        Chester○   Quaker Hill○ NAVAL
Guilford Deep River○   COAST GUARD ACADEMY SUBMARINE Pawcatuck○ ○Westerly
        ○Essex Ivoryton○   New London● Groton W. Mystic○ Stonington
        Old Saybrook ○Niantic   Poquonnock Mystic
Guilford   ○Old Lyme   Bridge
Madison Clinton   Morningside Pk. Watch Hill Pt.
hem Head Hammonasset Pt. Cornfield Pt.   Fishers I.

I S L A N D   The Race

O U N D   Plum I.
                    Orient Pt.
        Greenport   Gardiners
                    Bay   Gardiners I.
        Southold   Montauk Pt.

L O N G   I S L A N D   ATLANTIC
                              OCEAN   41°
Greenwich   30'   D   72°   E

CONNECTICUT — 127

DELAWARE

MILES
0 5 10 15 20

KILOMETERS
0 5 10 15 20

⊛ State Capital   ⊙ County Seats

PENNSYLVANIA

NEW JERSEY

NEW CASTLE

CHESAPEAKE BAY

Chester
Claymont
Bellefonte
Arden
Holly Oak
Wilmington
Winterthur
Elsmere
Minquadale
Centerville
Montchanin
Yorklyn
Westover Hills
Wilmington Manor
Newport
Christiana
Bear
New Castle
Marshallton
Newark
Brookside Park
Red Lion
St. Georges
Delaware City
Chesapeake & Delaware Canal
Odessa
Middletown
Townsend
Noxontown Lake
Blackbird
Clayton
Smyrna
Kenton
Hartly
Swedesboro
Penns Grove
Salem
Reeds
Port Penn
Bombay Hook Island
Goose Pt.
Kent I.
Kelly I.
Leipsic
Cheswold
Dupont Manor
Deepwater Pt.
Elmer
Monroeville
Newfield
Vineland
Millville
Bridgeton
Cedarville
Port Norris
Union L.
Elkton
North East
Cecilton
West Grove
Oxford
Pine Grove Res.
Rising Sun
Chestertown

Christina R.
Delaware R.
Salem R.
Maurice R.
Cohansey R.
Alloway Cr.
Smyrna R.
Leipsic R.
Duck Cr.
St. Jones R.
Chester R.
Sassafras R.
Bohemia R.
Elk River
Northeast R.
Susquehanna R.

442 ▲

76°   75°

30°

128 — DELAWARE

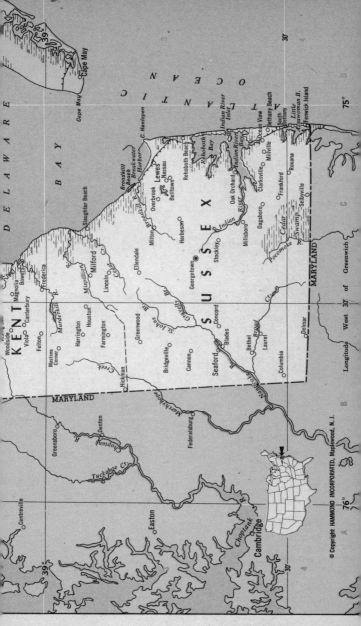

KENT

SUSSEX

MARYLAND

DELAWARE BAY

ATLANTIC OCEAN

Cape May
Cape May

C. Henlopen
Broadkill Beach
Breakwater Harbor
Lewes
Nassau
Rehoboth Beach
Indian River Inlet
Bethany Beach
South Bethany
Little Assawoman B.
Fenwick Island

Slaughter Beach
Overbrook
Belltown
Rehoboth Bay
Indian River Bay
Ocean View
Millville
Clarksville
Bethany

Milton
Harbeson
Oak Orchard
Frankford
Roxana

Frederica
Magnolia
Bowers
Viola
Canterbury
Ellendale
Stockley
Millsboro
Dagsboro
Selbyville

Woodside
Felton
Mastens Corner
Harrington
Houston
Greenwood
Georgetown
Concord
Farmington
Lincoln
Bridgeville
Cannon
Seaford
Bethel
Laurel
Delmar
Columbia
Blades
Hickman

Murderkill R.
Mispillion R.
Milford
Broadkill Cr.
Indian River
Cedar Swamp
Pocomoke R.
St. Johns Cr.
Nanticoke R.
Broad Cr.
Marshyhope Cr.

Greensboro
Denton
Federalsburg
Choptank
Tuckahoe Cr.
Centreville
Easton
Cambridge

Longitude West 30' of Greenwich

MARYLAND

© Copyright HAMMOND INCORPORATED, Maplewood, N.J.

DELAWARE — 129

FLORIDA

MILES
0 25 50 75

KILOMETERS
0 25 50 75

⊕ State Capital ⊛ County Seats

ATLANTIC OCE

FLORIDA — 131

Ft. Pierce
Port Salerno
Jensen Beach
Stuart
Hobe Sound
Lake Park
Jupiter
Indiantown
Riviera Beach
Palm Beach
W. Palm Beach
Lake Worth
Lantana
Boynton Beach
Delray Beach
Boca Raton
Deerfield Beach
Pompano Beach
Ft. Lauderdale
Hollywood
Miami Beach
Miami

OKEECHOBEE
Okeechobee
SEMINOLE IND. RES.
MARTIN
ST. LUCIE
PALM BEACH
BROWARD
Belle Glade
Pahokee
Canal Point
Clewiston
Moore Haven
HENDRY
GLADES
HIGHLANDS
L. Istokpoga
HARDEE
DESOTO
Arcadia
Nocatee
La Belle
Immokalee
Okaloacoochee Slough
BIG CYPRESS IND. RES.
SEMINOLE IND. RES.
COLLIER
Ochopee
Tamiami Trail
N. Miami
Coral City
Hialeah
Coral Gables
Miami
Biscayne Bay
Perrine
Homestead
Florida City
DADE
HOMESTEAD A.F.B.
BISCAYNE NAT'L PARK
Key Largo
Key Largo
Islamorada

Lake Okeechobee
BIG CYPRESS NAT'L PRESERVE
EVERGLADES NATIONAL PARK
Whitewater Bay
Florida Bay
C. Sable
Ponce de Leon B.

Charlotte Harbor
Punta Gorda
CHARLOTTE
Peace R.
MANATEE
Bradenton
Osamset
SARASOTA
Fruitville
Sunny land
Sarasota
Venice
Englewood
Pine I.
Cape Coral
Ft. Myers
Ft. Myers Beach
LEE
Bonita Springs
Naples
East Naples
Goodland
Ten Thousand Islands
Sanibel I.
Myakka R.

DE SOTO NAT'L MEM.

MEXICO

Clearwater
Belleair
Indian Rocks Beach
Madeira Beach
Treasure Island
St. Petersburg Beach
Largo
Pinellas Park
PINELLAS
Gulfport
Mullet Key
Safety Harbor
Sweetwater Cr.
Tampa
Riverview
Brandon
HILLSBOROUGH
MacDill A.F.B.
Ruskin
St. Petersburg
Tampa Bay

0   10 mi.
0   10 km.

Big Pine Key
Boca Chica Key
Sugarloaf Key
Key West
FLORIDA KEYS
Marathon
KEY WEST N.A.S.

Marquesas Keys

FT. JEFFERSON NAT'L MON., Dry Tortugas

Straits of Florida

© Copyright HAMMOND INCORPORATED, Maplewood, N.J.

Longitude West 82° of Greenwich

ALABAMA
Geneva
Graceville
JACKSON
Marianna
Chipola R.
Blountstown
HOLMES
Bonifay
Chipley
WASHINGTON
Southport
Ponce de Leon
Freeport
DeFuniak Springs
Crestview
Milligan
Baker
WALTON
OKALOOSA
Niceville
Valparaiso
Ft. Walton Beach
Destin
SANTA ROSA
Milton
Bagdad
Gulf Breeze
PENSACOLA
Warrington
W. Pensacola
Cantonment
Molino
Century
ESCAMBIA
Perdido
Perdido B.
Santa Rosa I.
SANTA ROSA ISLAND

Chattahoochee
GADSDEN
Quincy
LIBERTY
Bristol
Wewahitchka
GULF
Port St. Joe
C. San Blas
St. Joseph B.
CALHOUN
Blountstown
BAY
Panama City
Lynn Haven
Springfield
Parker
St. Andrew B.
Choctawhatchee R.
Choctawhatchee B.
St. Vincent I.

GULF OF MEXICO

WESTERN PART OF FLORIDA
Same scale as main map

86°   86°
84°
82° 30'
80°
26°
30°
A   B   C   D

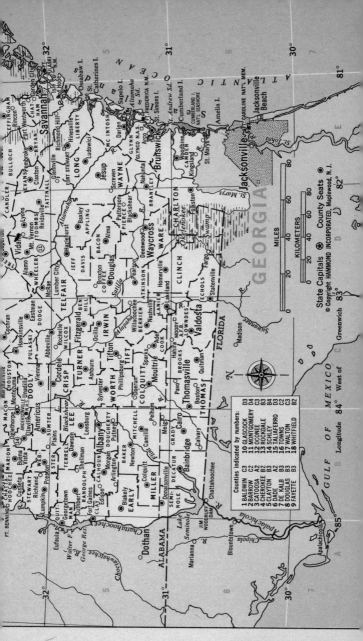

State Capitals ⊛ County Seats ◉
© Copyright HAMMOND INCORPORATED, Maplewood, N.J.

MILES
0   20   40   60   80
KILOMETERS
0   20   40   60   80

Counties indicated by numbers:			
1 BALDWIN	C3	10 GLASCOCK	D3
2 BARROW	C2	11 MONTGOMERY	D4
3 CATOOSA	A2	12 OCONEE	C3
4 CHEROKEE	B2	13 ROCKDALE	B3
5 CLAYTON	B3	14 SCHLEY	B4
6 DADE	A2	15 TALIAFERRO	C3
7 DE KALB	B3	16 TOWNS	C2
8 DOUGLAS	B3	17 WALTON	C3
9 FAYETTE	B3	18 WHITFIELD	B2

GEORGIA

ATLANTIC OCEAN

FLORIDA

ALABAMA

GULF OF MEXICO

PACIFIC

KAUAI

Kilauea
Hanalei
5,170
KAUAI
Kapaa
Nohili Pt.
Wailua
Kekaha
Kalaheo
Lihue
Waimea
Koloa
Hanapepe
Makahuena Pt.

Lehua
Puuwai
NIIHAU
C. Kawaihoa **COUNTY**

Kauai Channel

HONOLULU
Kahuku Pt.
Kahuku
Waialua
OA
Kaena Pt.
Wahiawa
Waianae
Kam
Waipahu
Aiea
Barbers Pt.
Ho
Pearl Har.
COUNTY

OAHU

0   5   10 mi.
0   5   10 km.

158° 15'   158°   21° 45'   157° 45'

Waialee
Kahuku Pt.
Kahuku
Mokuauia I.
Waimea
Laie
Laie Pt.
Kawailoa
Hauula
Mokuleia
Haleiwa
DILLINGHAM A.F.B.
Keana Pt.
Waialua
Kahana
Kahana Bay
Kaaawa
HONOLULU
Whitmore Vill.
Kaala 4,020
Schofield
Waikane
Kaneohe Bay
Barracks
Wahiawa
WHEELER A.F.B.
Makaha
Waipio Acres
Kahaluu
Kunia
Mokapu Pen.
Waianae
Pacific Palisades
Kaneohe
Mokapu
COUNTY
Kailua Bay
Maili
Pearl City
Kailua
Waipahu
Aiea
Maunawili
Nanakuli
Honouliuli
Waimanalo B
Pearl
Salt L.
Waimanalo
Makakilo City
Harbor
Waimanalo Be
Ewa
Hickam Housing
Mana
Barbers Pt. Housing
Iroquois Pt.
Honolulu
Woodlawn
Mak
Barbers Pt.
BARBERS PT. N.A.S.
Ewa Beach
Sand I.
Aina Haina
Koko Head
Mamala Bay
Waikiki
Kahala
Kuapa Pt.
Diamond Head
Maunalua Bay

21° 45'
21° 30'
21° 15'
158° 15'   158°   157° 45'

134 — HAWAII

157° E **156° F** **155° G**

180°   175°   170°   165°   160°   155°

Kure Atoll   Midway Is. *(U. S.)*
Pearl and Hermes Atoll

P A C I F I C

0 100 200 300 400 mi.
0   200   400 km.

H A W A I I A N

Lisianski I.

Laysan I.   Maro Reef

25°   'Gardner Pinnacles

O C E A N

French Frigate Shoals   Necker I.   Nihoa   **Tropic of Cancer**

Kauai

Niihau   Oahu   Molokai

Kaula   Lanai   Maui

I S L A N D S   Kahoolawe

Hawaii

KALAWAO COUNTY

Ilio Pt.   Kalaupapa   Halawa

naloa   Hoolehua

Kaunakakai   Pukoo   *Nakalele Pt.*

MOLOKAI   Wailuku   Kahului

Lahaina   Paia   **MAUI**

LANAI   Puunene

Lanai City   Makawao   *Kauiki Head*

Palaoa Pt.   Keokea   Hana

Molokini   10,023   HALEAKALA NAT'L PARK

KAHOOLAWE

Kealaikahiki Pt.

C O U N T Y

Alenuihaha Channel

O C E A N

N

Upolu Pt.

Hawi   Kapaau (Kohala)

Kawaihae   Haina   Honokaa   **HAWAII**

Kawaihae Bay   Paauilo

PUUKOHOLA HEIAU   Ookala

N.H.S.

Waikii   Hakalau

Mauna Kea   Pepeekeo

13,796

Keahole Pt.   KALOKO-HONOKOHAU   Papaikou

Kailua   NAT'L HIST. PARK   **Hilo**

(Kailua Kona)   Keaau

Holualoa   Kurtistown   Kapoho

Kealakekua   Mountainview   *C. Kumukahi*

Captain Cook   *Mauna Loa*   Pahoa

PUUHONUA O HONAUNAU   13,67   *Kilauea Crater*

NAT'L HIST. PARK   Kalapana

HAWAII VOLCANOES

HAWAII Milolii   NAT'L PARK   **COUNTY**

Pahala

Naalehu

Ka Lae (South Cape)

HAWAII

MILES

0 10 20 30 40 50 60

KILOMETRES

0 10 20 30 40 50 60

State Capital   ✪

© Copyright HAMMOND INCORPORATED, Maplewood, N.J.

gitude **157°** West of E   Greenwich **156°** F   **155°** G

22° 1

25°

20° 2

21°

3

20°

19°

5

IDAHO

State Capitals ⊕

MILES
20 40 60 80

KILOMETRES
20 40 60 80

N

112°
ALBERTA
BLACKFEET
INDIAN
RESERVATION

CONTINENTAL

DIVIDE

Great Falls

Smith R.

Missouri R.

Helena ⊕

Butte

Canyon
Ferry
Res.

R O C K Y   M O U N T A I N S

CONTINENTAL   DIVIDE

114°
WATERTON-GLACIER
WATERTON LAKES
NAT'L PARK
INT'L PEACE PK.
GLACIER
NATIONAL
PARK

Hungry
Horse
Res.

Flathead

Kalispell

Flathead L.

FLATHEAD
INDIAN
RESERVATION

Flathead
R.

Clark Fork

Blackfoot R.

Clark Fork

Missoula

Bitterroot R.

116°
BRITISH COLUMBIA

Lake
Koocanusa

Kootenai R.

Libby

MONTANA

Lolo Pass
5,187

B I T T E R R O O T

Clark Fork

C A B I N E T   M T S.

BOUNDARY
Metaline Falls

KALISPEL
IND. RES.

Bonners Ferry

Priest
Lake

Priest River

Clark
Fork

Pend
Oreille

Sandpoint

Pend Oreille
Lake

BONNER

Spirit Lake

Hayden
Rathdrum
Post Falls

Spokane R.

Spokane

WASHINGTON

Pullman

Colfax

Coeur d'Alene

Coeur
d'Alene L.

KOOTENAI

Smelterville
Osburn
Kellogg
Wallace
Mullan

Avery

SHOSHONE

St. Joe R.

St. Maries

BENEWAH

Potlatch

LATAH

Troy
Moscow
Genesee

Lapwai

NEZ PERCE
NAT'L HIST. PARK

Lewiston

Snake River

E. Sister Pk.
6,866

Elk River

CLEARWATER

Headquarters

Pierce

Orofino

Weippe

Kamiah

Kooskia

Lochsa R.

Selway R.

C L E A R W A T E R

Lolo
Fork

I D A H O

LEWIS

Craigmont
Nezperce
Cottonwood

Grangeville
White
Bird

Kooskia

NEZ PERCE

48°
46°

BOUNDARY

136 — IDAHO

138 — ILLINOIS

ILLINOIS — 139

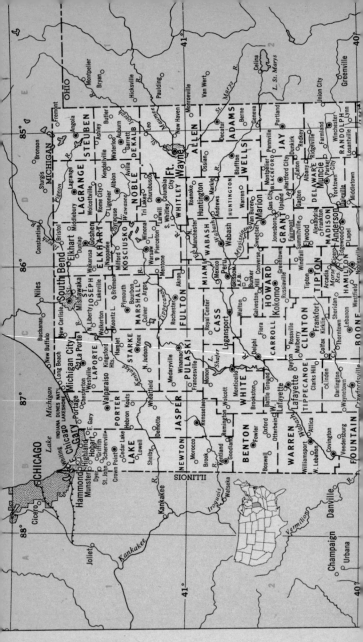

IOWA

MILES
0 20 40 60

KILOMETERS
0 20 40 60

State Capital ⊛
County Seats ⊙

Austin

Caledonia
New Albin
Lansing
MITCHELL   Cresco   WINNESHIEK   Waukon
St. Ansgar   Decorah
Riceville   ALLAMAKEE
Osage   HOWARD   Elma
Nora   Calmar   Postville   EFFIGY MOUNDS
Sprs.   NAT'L MON.
FLOYD   Charles   New   Monona   McGregor   Wisconsin R.
Rockford   City   Hampton   43°
CHICKASAW   Nashua   W. Union   CLAYTON   Elkader
Greene   Sumner   Fayette   Guttenberg
Allison   Clarksville   Tripoli   FAYETTE   Strawberry Pt.
BUTLER   Waverly   Oelwein   Lancaster
Shell Rock   BREMER   Denver   Edgewood   Platteville   Darlington
Parkersburg   Hazleton   DELAWARE   WISCONSIN
            BLACK HAWK   BUCHANAN   Dyersville   Dubuque   ILLINOIS
Cedar Falls   Jesup   Manchester   Farley   Galena
GRUNDY Waterloo   Evansdale   Independence   Hopkinton   DUBUQUE
Grundy   Hudson   Washburn   Cascade   Bellevue
Ctr.   Reinbeck   La Porte City   Central   RIVER
Conrad   Center   City   Monticello   JACKSON
Gladbrook   Traer   Point   Anamosa   Maquoketa
TAMA   Vinton   LINN   JONES   Preston   Sabula
Marshalltown   BENTON   Marion   Wyoming   42°
    Toledo   Hiawatha   CLINTON
HALL   SAC & FOX   Tama R.   Cedar Rapids   Olin   Clinton
    IND. RES.   Belle Plaine   Lisbon   De Witt   Camanche
ER   Grinnell   Amana   Mt. Vernon   Mechanics   Clarence
    Brooklyn   Marengo   ville   Tipton   SCOTT
POWESHIEK   Victor   IOWA   Iowa   CEDAR   Durant
    Williamsburg   Coralville   City   W. Branch   Le Claire   Bettendorf
    Montezuma   N. English   University Hts.   H. HOOVER   Davenport   Moline
Pella   JOHNSON   NAT'L HIST. SITE   Wilton   Rock Island
MAHASKA   New Sharon   Wellman   W. Liberty   MUSCATINE
Oskaloosa   What Cheer   Kalona   Muscatine   ILLINOIS
    Sigourney   Keota   WASHINGTON   Aledo
N   KEOKUK   Washington   Columbus
    Eddyville   Hedrick   Brighton   Jct.   LOUISA
Lovilia   WAPELLO   Winfield   Wapello   Morning Sun
MONROE   Ottumwa   JEFFERSON   HENRY   Mediapolis
    Albia   Fairfield   Monmouth   Galesburg
            Eldon   Mt. Pleasant   New London   DES MOINES   41°
    Moravia   DAVIS   VAN BUREN   W. Burlington   RIVER
Mystic   Centerville   Bloomfield   Milton   Burlington
PANOOSE   Moulton   Keosauqua   West Point
Unionville   Farmington   Ft. Madison
            Kahoka   LEE   Keokuk   MISSISSIPPI

3° West of E Greenwich   92°   91°   G

© Copyright HAMMOND INCORPORATED, Maplewood, N. J.

IOWA — 143

KANSAS

102° A 101° B 100° C 99°

1

40°

Swanson L.
McCook
Republican R.
Alma
Harlan Co. Lake

CHEYENNE
St. Francis
Bird City

Atwood
Ct.
Oberlin
Sappa
Norton Res.
Almena
Norton
PHILLIPS
Kensington
Phillipsburg
Logan
SMITH
Le
Smi

RAWLINS
DECATUR
Lenora
NORTON
Fork
Kirwin Res.

SHERMAN
Goodland
Colby
North
Hoxie
Hill City
GRAHAM
Webster Res.
Stockton
ROOKS
South
Fork
Palco
Plainville
Osborne
OSBOR
HILLS

THOMAS
SHERIDAN
SMOKY
Natoma

2

39°

Mt. Sunflower 4,039
LOGAN
Oakley
Quinter
WaKeeney
Saline R.
Russ
RUSSE

Sharon Sprs.
Russell Sprs.
Gove
GOVE
TREGO
Cedar Bluff Res.
Ellis
ELLIS
Hays
Victoria

WALLACE
Smoky
Hill
River

Tribune
Leoti
Scott City
LANE
Dighton
NESS
Ness City
La Crosse
Hoisington
Cheyenne Bottoms
BART

GREELEY
WICHITA
SCOTT
RUSH
Great Bend

3

38°

KEARNY
L. McKinney
Lakin
FINNEY
Garden City
Pawnee R.
HODGEMAN
Jetmore
Larned
FT. LARNED NAT'L. HIST. SITE
PAWNEE
St. John
STAFF

Syracuse
Arkansas
R.
Cimarron
Kinsley
Macksville
Staff

HAMILTON
Dodge City
Ft. Dodge
Spearville
EDWARDS
Pratt
PRAT

Johnson
Ulysses
HASKELL
GRAY
Montezuma
FORD
Greensburg
Haviland

STANTON
GRANT
Sublette
Bucklin
KIOWA
Satanta
Minneola
Coldwater

4

37°

Richfield
R.
Fowler
Plains
Meade
Ashland
Protection
BARE

MORTON
STEVENS
Hugoton
SEWARD
MEADE
CLARK
COMANCHE
Medicine

Cimarron
R.
Elkhart
Liberal
Cr.

KANSAS

MILES
0 20 40 60 80

KILOMETERS
0 20 40 60 80

TEXAS

State Capital ✪   County Seats ●

© Copyright HAMMOND INCORPORATED, Maplewood, N.J.

Beaver
Cimarron R.
Alva
R.

indi
1 GEAF
2 JEFF
3 LEAV
4 SHAV
5 WYA

102° A 101° B 100° C 99°

KANSAS — 145

WESTERN PART OF KENTUCKY
Same scale as main map

W. Frankfort

Marion

Harrisburg

Shawneetown

Uniontown

Morganfield
UNION
Sturgis

Greenfield

Connersville

Rushville

Shelbyville

Anna

Vienna

Rosiclare

Golconda

Clay

CRITTENDEN

Flatrock

Greensburg

Columbus

Bate

Cape Girardeau

Mound City

La Center

Barlow

Metropolis

Lone Oak

PADUCAH

Woodlawn

Salem

LIVING-
STON

Smithland

Marion

Calvert
City

Eddyville

CALDWELL

Princeton

KENTUCKY DAM

N. Vernon

Austin

Scottsburg

Madison

Carrollton

Bedford

TRIMBLE

ILLINOIS

Cairo

Wickliffe

McCRACKEN

Benton

Kentucky Lake

Cadiz

TRIGG

Charleston

Sikeston

E. Prairie

Bardwell

CARLISLE
Arlington

MARSHALL

LAND BETWEEN

New
Albany

New Madrid

Clinton

HICKMAN

Mayfield

GRAVES

CALLOWAY

Murray

THE LAKES

REC. AREA

St. Matthews

LaGrange
OLDHAM
Crestwood

HEN
Em

FULTON

Fulton

TENNESSEE

New
Albany

Louisville

Shively

Middletown
Prospect

SHELB

Boonville

Pleasure Ridge Pk.

Valley Sta.

Buechel

Okolona

Taylorsvi

SHELB

Mt. Vernon

Ohio

Evansville

River

INDIANA

MEADE

Brandenburg

Mt. Washington

Shepherdsville

Sol

4

8

Henderson

Hawesville

Tell City

Irvington

Muldraugh

Radcliff

FT.
KNOX

BULLITT

Bloomfield

Uniontown
HENDERSON

Morganfield

Cloverport

Hardinsburg

Vine Grove

Lebanon
Jct.

NELSON

Bardstown

New Haven

WASHING

Owensboro

Whitesville

Fordsville

BRECKINRIDGE

Rough R.
L.

HARDIN

Elizabethtown

6

Springfie

Leba

UNION
Sturgis

DAVIESS

Calhoun

Livermore

OHIO

Rough

Hodgenville

Rollins

Fork

MARIO

Sebree

Green R.

Hartford

Leitchfield

ABRAHAM LINCOLN BIRTHPLACE
NAT'L. HIST. SITE

WEBSTER

Dixon

McLEAN

Beaver Dam

GRAYSON

HART

Greens-
burg

Campbellsville

Clay

Providence

Central City

Nolin
L.

Green

Munfordville

TAYLOR

HOPKINS

Earlington

Madisonville

Mortons Gap

Drakesboro

Morgantown

EDMONSON

Horse Cave

GREEN

Green R.
Res.

Columbia

ADAIR

CALDWELL

Dawson Sprs.

Nortonville

Greenville

MUHLENBERG

BUTLER

Brownsville

MAMMOTH CAVE
NAT'L. PK.

Cave City

Princeton

Crofton

Lewisburg

Hadley

BARREN

Jametow
R
Alb

TRIGG
Lake

Cadiz

CHRISTIAN

Hopkinsville

LOGAN

Bowling Green

Smiths
Grove

Glasgow

Edmonton

METCALFE

WOLF CR. DA

2

Barkley

Pembroke

Elkton

Russellville

Auburn

WARREN

Barren R.
Lake

Scottsville

MONROE

Burkesville

Cumberlan

FT. CAMPBELL

TODD

Guthrie

Adairville

SIMPSON

Franklin

ALLEN

Tompkinsville

Clarksville

TENNESSEE

Portland

Springfield

Gamaliel

---

146 — KENTUCKY

KENTUCKY

MILES
0 5 10   20   30   40   50 60

KILOMETERS
0 5 10 20 30 40 50 60

⊗ State Capital
⊚ County Seats

Counties indicated by numbers:
1 CAMPBELL D2   6 LARUE   C3
2 CUMBERLAND C4   7 ROBERTSON D2
3 GARRARD   D3   8 SPENCER   C2
4 JEFFERSON C2   9 WOODFORD   D2
5 JESSAMINE D3

© Copyright HAMMOND INCORPORATED, Maplewood, N.J.

LOUISIANA — 149

150 — MAINE

MAINE

State and Provincial Capitals ⊛
County Seats ◉

© Copyright HAMMOND INCORPORATED, Maplewood, N.J.

MILES
0   10   20   30   40   50

KILOMETERS
0 10 20 30 40 50

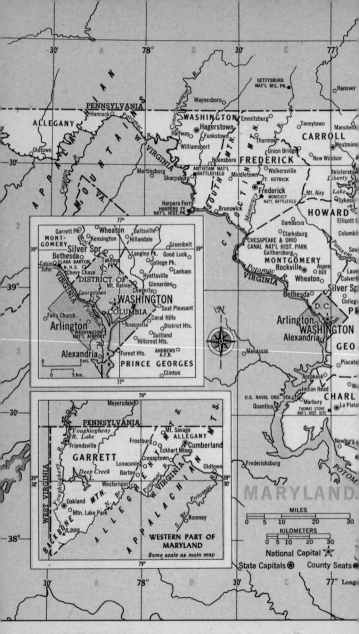

MARYLAND — 153

154 — MASSACHUSETTS

MASSACHUSETTS

MILES
0   10   20   30

KILOMETERS
0   10   20   30

● State Capitals
◎ County Seats (Shire Towns)

MASSACHUSETTS BAY

Cape Cod
Provincetown

CAPE COD
NAT'L
SEASHORE

Wellfleet

Cape Cod

Bay

Orleans

Cape Cod Canal
Sandwich BARNSTABLE
Barnstable   Harwich   Chatham
Buzzards   Hyannis
Bay   Osterville S.
OTIS   Yarmouth
A.F.B.
                                Monomoy Pt.

Nantucket

Sound

Great Pt.

Muskeget Chan. NANTUCKET

Nantucket
Nantucket I.

ATLANTIC   OCEAN

© Copyright HAMMOND INCORPORATED, Maplewood, N.J.   41°

West of   Greenwich 71°   30'   70°

MICHIGAN

© Copyright HAMMOND INCORPORATED, Maplewood, N.J.

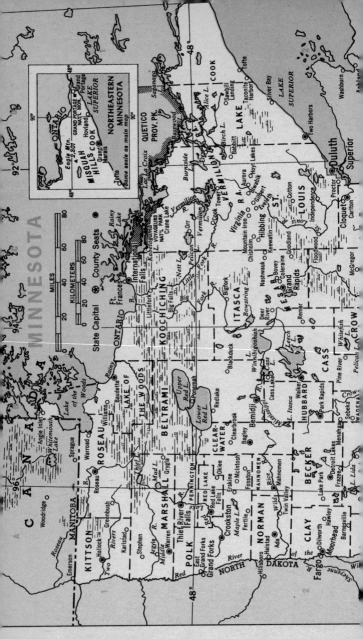

MINNESOTA — 159

162 — MISSOURI

MISSOURI — 163

MILES
0 10 20 40 60 80 100
KILOMETERS
0 20 40 60 80 100

State Capital ⊛   County Seats ◉

MONTANA

MONTANA — 165

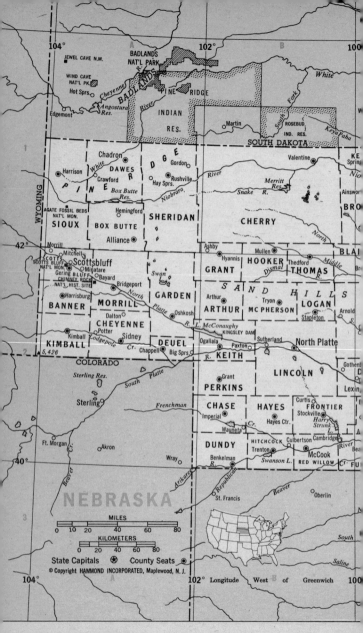

104°   A   102°   B   100

JEWEL CAVE N.M. ■

BADLANDS
NAT'L PARK

WIND CAVE
NAT'L PK.

Hot Sprs. ●   BADLANDS   White

Cheyenne   PINE RIDGE
Edgemont ●   Angostura   INDIAN   South Fork
Res.   River   RES.

Martin ●   ROSEBUD
IND. RES.   Keya Paha

SOUTH DAKOTA   KE
Spring

1   Chadron ●   R   Gordon ●   Valentine ●
Harrison ●   DAWES   I   Rushville ●   Nio
Crawford ●   D   Hay Sprs. ●   River
White   G   Niobrara   Merritt   Ainswort
Box Butte   E   Snake R.   Res.   BRO
Res.

P   Hemingford ●
I   SHERIDAN   CHERRY
SIOUX   N   North
BOX BUTTE   E   42°
Agate Fossil Beds   Alliance ●
Nat'l Mon.

42°   Morrill ●   Ashby ●   Mullen ●   BLAI
Mitchell ●   Hyannis ●   HOOKER   Thedford ●   Br
SCOTTS BLUFF   Scottsbluff ●   GRANT   Dismal   THOMAS
Nat'l Mon.   Minatare ●   S   A   N   D   Middle
Gering ●   Bayard ●   Swan   Arthur ●   Tryon ●   H   I   L   L   S
CHIMNEY ROCK   Bridgeport ●   GARDEN   McCONAUGHY   LOGAN
NAT'L HIST. SITE   North   ARTHUR   McPHERSON   Stapleton ●   Arnold ●

BANNER   MORRILL   Platte   Oshkosh ●   L. McConaughy
Dalton ●   KINGSLEY DAM   Sutherland ●
CHEYENNE   R.   Ogallala ●   Paxton ●   North Platte ●
Potter ●   Sidney ●   DEUEL   LINCOLN
KIMBALL   Lodgepole   Cr.   Chappell ●   Big Sprs. ●   R.   KEITH   Gothen
Kimball ●   △ 5,426   2   Gering ●
COLORADO   Grant ●   Lexin

Sterling Res.   South   PERKINS   Curtis ●   FRONTIER
Sterling ●   Platte   CHASE   HAYES   Stockville ●   Harry
Frenchman   Imperial ●   Curtis ●   Strunk   D
Wauneta ●   Cr.   Hayes Ctr. ●   L.
Ft. Morgan ●   Akron ●   HITCHCOCK   Culbertson ●   Cambridge ●   River   Bea
DUNDY   Trenton ●   McCook ●   FU
Wray ●   Benkelman ●   Swanson L.   RED WILLOW   Cr.
R.   Republican   Oberlin ●

NEBRASKA   St. Francis ●   Beaver   South

3   MILES   Saline

0 10 20   40   60   80

KILOMETERS
0   20   40   60   80

★ State Capitals   ✪ County Seats ●
© Copyright HAMMOND INCORPORATED, Maplewood, N.J.

104°   A   102° Longitude   West   of   Greenwich   100

WYOMING

166 — NEBRASKA

NEBRASKA — 167

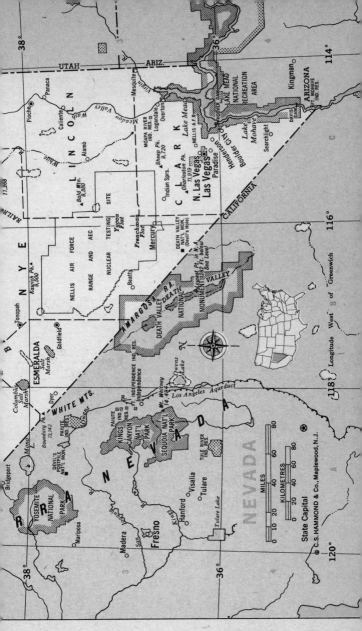

NEW HAMPSHIRE

MILES
0　5　10　15　20　25　30　35

KILOMETRES
0　5　10　15　20　25　30　35

State Capitals ⊛

45

NEW JERSEY — 173

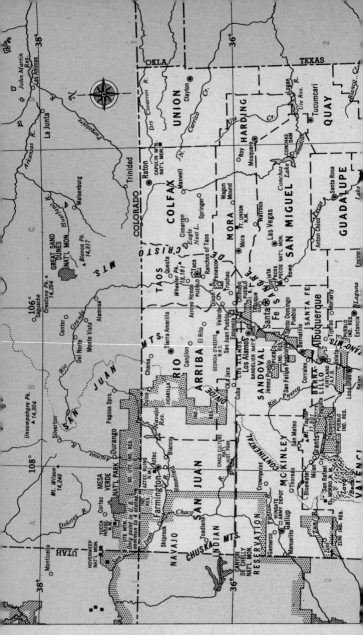

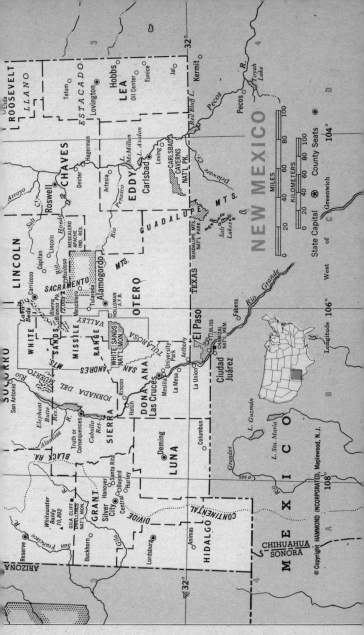

NEW MEXICO

State Capital ⊛ County Seats ⦿

MILES
0 20 40 60 80 100

KILOMETERS
0 20 40 60 80 100

West of Greenwich

Longitude 106°

© Copyright HAMMOND INCORPORATED, Maplewood, N.J.

NORTHERN PART OF
NEW YORK
Same scale as main map

176 — NEW YORK

NEW YORK — 177

178 — NORTH CAROLINA

NORTH CAROLINA — 179

NORTH DAKOTA — 181

© Copyright HAMMOND INCORPORATED, Maplewood, N.J.

Longitude 98° West

OKLAHOMA — 185

186 — OREGON

OREGON

120°   118°

C   D   E

Snake   R.   ○ Pomeroy   Lewiston

HANFORD
ATOMIC
ENERGY
RES.

Sacajawea

Richland   Lake   Dayton ○

Pasco

L. Wallula   WHITMAN   Walla Walla
MISSION
NAT'L
HIST. SITE   46°

○ Milton-Freewater

McNARY DAM   ○ Athena   ● Weston   Elgin ○   Wallowa ○

Columbia   R.   Hermiston   Pendleton   WALLOWA

● Capital   ⊛   ● County Seats

WASHINGTON

JOHN
DAY DAM   Wasco
Dalles   Moro

Arlington   Umatilla   R.

MORROW   Pilot Rock   UMATILLA
IND. RES.

UMATILLA   La Grande ●   HELLS CANYON
NAT'L REC.
AREA

○ Joseph   Enterprise ●

GILLIAM   Condon   Heppner ○   UNION   Union ○   WALLOWA
MTS.

SHERMAN

CO   John Day Fossil
Beds Nat'l Mon.   Fossil   Kinzua ○   North   Fk.   BLUE   Halfway ○   Powder   R.

JEFFERSON   John Day   Day   JOHN DAY FOSSIL
BEDS NAT'L MON.   GRANT   Bates ○   BAKER   ● Baker

Madras ●   WHEELER   John Day ●   Prairie City ●

Prineville ●   CROOK   Mt. Vernon   Canyon City ●   Huntington ○   ○ Weiser

Strawberry
Mtn.
9,038

Prineville Res.   Crooked   R.   Vale ●   ○ Payette   44°

HIGH   DESERT   Ontario ○   Payette   R.

CHUTES   Nyssa ○   Boise   R.

Burns ●   Silver   Warm   Sprs. Res.   Malheur   R.   ○ Nampa

Hines ○   HARNEY   BASIN   MALHEUR   L. Owyhee

Silver L.   Silver   Malheur
L.   LAVA   BEDS

LAKE   Silver L.   C.   Harney L.

Silver L. ○   SAND   HARNEY   Jordan
Valley ○

DUNES   STEENS   Antelope
Res.

L. Abert   VALLEY   Bluejoint L.   Owyhee   IDAHO

Flagstaff L.   ALVORD
DESERT

Hart L.   Alvord L.

Drews
Res.   Lakeview ●   Crump L.   WARNER

Clear Lake
Res.   Goose L.   NEVADA   42°

○ McDermott

© Copyright HAMMOND INCORPORATED, Maplewood, N.J.

of   Greenwich   120°   118°

LAKE ERIE

NEW YORK

PITTSBURGH area inset:
Ohio, O. W. View, Oakmont, Sharpsburg, Etna, Millvale, Bellevue, Avalon, McKees Rocks, PITTSBURGH, ALLEGHENY, Wilkinsburg, Monroeville, Plum, Crafton, Carnegie, Dormont, Swissvale, Munhall, Turtle Cr., Braddock, Braddock, Trafford, Baldwin, Whitehall, Duquesne, Mt. Lebanon, Brentwood, McKeesport, Bridgeville, W. Mifflin, Bethel Pk., White Oak

ERIE

Lawrence Pk., North East, Wesleyville, Lake City, Fairview, Erie, Girard, Waterford, Albion, Edinboro, Union City, Corry, Warren, Bradford, Eldred, Shinglehouse, Port Allegany, Coudersport, Conneautville, Saegertown, Cambridge Sprs., Youngsville, McKEAN, Smethport, Mt. Jewett, Linesville, CRAWFORD, Meadville, Titusville, Sheffield, Kane, Pymatuning Res., Cochranton, VENANGO, Oil City, FOREST, Tionesta, Johnsonburg, St. Marys, CAMERON, Emporium, Greenville, Franklin, Polk, Ridgway, Brockway, ELK, Shenango R. Lake, MERCER, Mercer, Grove City, CLARION, Knox, Clarion, Brockport, JEFFERSON, Brookville, Du Bois, Sandy, CLEARFIELD, Farrell, Sharon, Wheatland, New Wilmington, Slippery Rock, Rimersburg, New Bethlehem, Reynoldsville, Sykesville, Clearfield, Youngstown, NEW CASTLE, Bessemer, LAWRENCE, Chicora, E. Brady, Punxsutawney, Curwensville, CEN, Ellwood City, BUTLER, Lyndora, Butler, Kittanning, Mahoning Cr., Philipsburg, Pleasant Gap, Beaver Falls, BEAVER, Zelienople, Ford City, Houtzdale, Sta. Co., Beaver, New Brighton, Freeport, ARMSTRONG, INDIANA, Clymer, Barnesboro, Tyrone, Midland, Monaca, Economy, Natrona Hts., Leechburg, Patton, Bellwood, Aliquippa, Ambridge, Arnold, Vandergrift, Indiana, Spangler, CAMBRIA, Altoona, Sewickley, Coraopolis, Glenshaw, Lower Burrell, Homer City, Ebensburg, Gallitzin, Huntingd, Coraopolis, ALLEGHENY, New Kensington, Nanty Glo, Portage, Hollidaysburg, PITTSBURGH, Plum, Blairsville, Johnstown, Roaring Spr., Mt. Union, McDonald, Trafford, Jeannette, Derry, Westmont, Geistown, Claysburg, Canonsburg, McKeesport, Irwin, Greensburg, Latrobe, Ligonier, Windber, BEDFORD, Washington, Clairton, New Kensington, Youngwood, Boswell, HUNTINGDON, Monongahela, Donora, WESTMORELAND, Mt. Pleasant, Charleroi, Bentleyville, Monessen, Scottdale, SOMERSET, Central City, WASHINGTON, California, Connellsville, Somerset, Bedford, Everett, FULTON, Centerville, FAYETTE, Oliver, Rockwood, Berlin, McConnellsburg, Chamb, Waynesburg, Masontown, Uniontown, FR, GREENE, Fairchance, FT. NECESSITY NAT'L BATTLEFIELD, Meyersdale, Mercersburg, Greencas, Bobtown, FRIENDSHIP HILL N.H.S., Youghiogheny, Mt. Davis 3,213, Hyndman, WEST VIRGINIA, River Lake, Frostburg, Cumberland, Morgantown, Potomac

WEST VIRGINIA

OHIO

188 — PENNSYLVANIA

PENNSYLVANIA

MILES
0 10 20 30 40 50

KILOMETERS
0 10 20 30 40 50

State Capitals ⊛
County Seats ⊛

BUCKS
MONT-GOMERY

Willow Grove · Abington
Bridgeport · Jenkintown   Bristol
Conshohocken
Cheltenham
Haverford   PHILADELPHIA
Ardmore   N.J.   Delaware R.   Willingboro   40°
Newtown Sq.
Darby   Pennsauken
Lansdowne
Yeadon   Camden
Prospect Pk.   Gloucester City

5 mi.
5 km.
75°

Pepacton Res.
NEW YORK   42°
Hallstead   N
Elkland   Susquehanna
Athens · Sayre   New Milford
Mansfield   BRADFORD   Montrose   Delaware R.
Troy · Towanda   WAYNE   Mt. Ararat 2,667
TIOGA   SUSQUEHANNA
Blossburg   Canton   Forest City   Honesdale
SULLIVAN   Carbondale   Hawley   Port Jervis
Tunkhannock   Blakely   L. Wallenpaupack   2
WYOMING   Dickson City · Winton   Milford   Matamoras
Williamsport   Laporte   Old Forge · Dunmore   DELAWARE WATER GAP NAT'L REC. AREA
Montoursville   LUZERNE   Scranton   PIKE
Muncy   Kingston   Pittston   LACKA-WANNA   Newton
Montgomery   Plymouth   Plains·   Wilkes-Barre   POCONO MTS.
Lock Haven   Watsontown   Glen Lyon   Nanticoke
Mill Hall   COLUMBIA   Berwick   Lehigh   41°
Milton   Espy   Nescopeck   Freeland   E. Stroudsburg   Delaware Water Gap
Lewisburg   Danville   Bloomsburg   Hazleton   CARBON   Stroudsburg
Mifflinburg   MONTOUR   McAdoo   Weatherly   MONROE   Bangor
UNION   NORTH-   Mt.   Shenandoah   Jim Thorpe   Pen Argyl
Middleburg   Sunbury   UMBERLAND   Carmel   Lansford   Lehighton   NORTHAMPTON
Selinsgrove   Shamokin   Kulpmont   Frackville   Tamaqua   Palmerton   Nazareth   Phillipsburg
SNYDER   Minersville   Northampton   Wilson·   Easton
Mifflintown   Williamstown   Schuylkill Haven   Pottsville   Catasauqua   Bethlehem   Somerville
Newport   Millersburg   Lykens   Pine Grove   Hamburg   Allentown   LEHIGH   Hellertown
Bloomfield   SCHUYLKILL   Northampton   Emmaus
DAUPHIN   Myerstown   BERKS   Kutztown   Quakertown
Duncannon   Marysville   Annville   Fleetwood   Perkasie   BUCKS   Trenton
Harrisburg   Hershey   LEBANON   Wyomissing   Boyertown   Souderton   Doylestown   Morrisville
Camp Hill   Steelton   Palmyra   Lebanon   Reading   Laureldale   Lansdale   Fairless Hills
Carlisle   New   Middletown   Shillington   Birdsboro   Quakertown   MONTGOMERY   Hatboro   Abington   Levittown
Mechanicsburg   Cumberland   Manheim   Spring City   Pottstown   Royersford   Norristown   Bristol
Mt. Holly Sprs.   Elizabethtown   Mt. Joy   Lititz   New Holland   Phoenixville   Conshohocken   PHILADELPHIA
Marietta   Ephrata   Downingtown   CHESTER   Bryn Mawr   PHILADELPHIA   40°
Columbia   LANCASTER   W.   Upper Darby   Camden
Wrightsville   Lancaster   Coatesville   Chester   DELAWARE   Media   Woodbury
York   Millersville   Parkesburg   Kennett Sq.   Chester
ADAMS   Dallastown · Red Lion   Quarryville   Oxford   Marcus Hook
Gettysburg   Hanover   YORK   DEL.   Wilmington
Littlestown   Parkville · New Freedom   CONOWINGO DAM   Elkton
MARYLAND   CONOWINGO DAM

© Copyright HAMMOND INCORPORATED, Maplewood, N.J.
Longitude West 77° of Greenwich

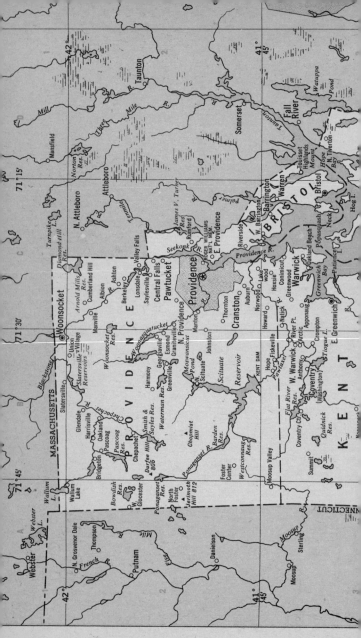

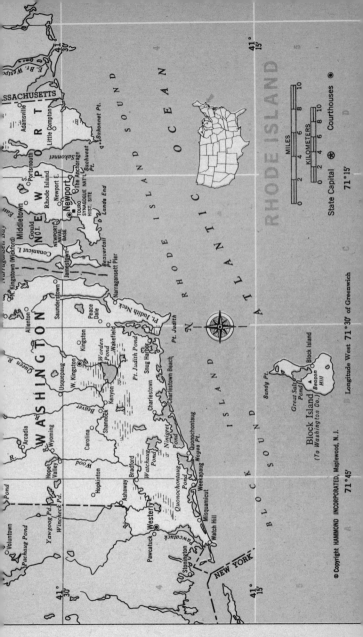

MASSACHUSETTS

Adamsville
Little Compton

E Br. Westport

S. Portsmouth
Rhode Island
Newport E.
Newport
TOURO
SYNAGOGUE NAT'L
HIST. SITE
The Anchorage
NEWPORT
NAVAL
BASE

NE W P O R T

Middletown
Gould I.
NO. 1

Conanicut I.

Jamestown

N. Kingstown (Wickford)

Saunderstown

Allenton

Exeter

W A S H I N G T O N

Queen R.

Arcadia

Wyoming

Hope
Valley

Hopkinton

Ashaway

Pawcatuck

Westerly

Watch Hill

Pachaug Pond

Voluntown

Yawgoog Pd.

Wincheck Pd.

Pond

Beaver

Carolina

Shannock

Kenyon

Bradford

Wood R.

Watchaug
Pond

Quonochontaug
Pond

Quonochontaug

Misquamicut

Pawcatuck R.

Stonington

NEW YORK

Usquepaug

W. Kingston

Kingston

Peace
Dale

Worden
Pond

Wakefield

Pt. Judith Pond

Snug Harbor

Charlestown

Charlestown Beach

Ninigret
Pond

Weekapaug Pt.

Noyes Pt.

Pt. Judith Neck

Pt. Judith

Narragansett Pier

Bonnet
Pt.

Beavertail
Pt.

Sachuest
Pt.

Sakonnet Pt.

Lands End

Sakonnet

Narragansett Bay

East

R H O D E   I S L A N D   S O U N D

A T L A N T I C   O C E A N

B L O C K   I S L A N D   S O U N D

Sandy Pt.

Great Salt
Pond

Block Island

Beacon
Hill

Block Island
(To Washington Co.)

RHODE ISLAND

State Capital ⊛   Courthouses ◉

MILES
0 2 4 6 8 10

KILOMETERS
0 2 4 6 8 10

41°
15'

71°15'

Longitude West 71°30' of Greenwich

71°45'

41°
30'

41°
15'

41°
30'

© Copyright HAMMOND INCORPORATED, Maplewood, N.J.

SOUTH CAROLINA

MILES
0   20   40   60

KILOMETERS
0   20   40   60

⊛ State Capital
◉ County Seats

SOUTH CAROLINA — 193

194 — SOUTH DAKOTA

SOUTH DAKOTA — 195

196 — TENNESSEE

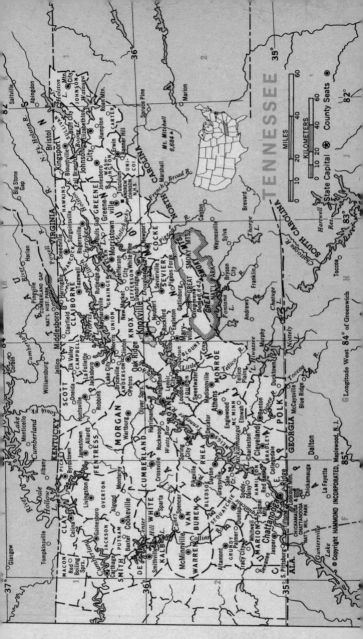

TEXAS

GULF OF MEXICO

MEXICO

NEW MEXICO

CHIHUAHUA

COAHUILA

TAMAULIPAS

NUEVO LEÓN

28°
26°
30°

SCALE
MILES
0 40 80 120
KILOMETERS
0 40 80 120

⊛ State Capital ⊛ County Seats

HOUSTON inset

HARRIS

L. Houston
Dayton
Liberty
LIBERTY
Spring Valley
Hunters Cr. Village
W. Univ. Pl.
Bellaire
Barrett
Crosby
Highlands
Chambers
Baytown
Channelview
HOUSTON
San Jacinto
Galena Pk.
Jacinto City
Pasadena
La Porte
CHAMBERS
JOHNSON SPACE CTR.
Sugar Land
Richmond
Rosenberg
FORT BEND
W. Columbia
Sugar Lang
League City
Dickinson
Hitchcock
La Marque
Alvin
Texas City
GALVESTON
Galveston
BRAZORIA
Angleton
Brazos
Bay
GALVESTON
GULF OF MEXICO

95°
94°
29°30'
96°

MILES
0 10 20 30

© Copyright HAMMOND INCORPORATED, Maplewood, N.J.

Western Part of Texas inset

NEW MEXICO

CHIHUAHUA

CARLSBAD CAVERNS NAT'L PK.
GUADALUPE MTS. NAT'L PARK
Guadalupe Pk. 8,749'
Red Bluff
Mentone
LOVING
Pecos
Toyah
REEVES
DAVIS MTS.
Mt. Livermore 8,382'
Ft. Davis
FT. DAVIS
Alpine
BREWSTER
Marfa
PRESIDIO
Presidio
Ojinaga
BIG BEND NAT'L PARK
Van Horn
CULBERSON
HUDSPETH
Sierra Blanca
SIERRA DIABLO MTS.
FT. BLISS
BIGGS A.F.B.
EL PASO
El Paso
Ciudad Juárez
Villa Ahumada
Moctezuma

WESTERN PART OF TEXAS
Same scale as main map

106°
104°
102°
32°
30°
28°
26°

R i o G r a n d e
Rio Conchos

M E X I C O

Main map place names

Nederland
Groves
Port Arthur
Anahuac
HOUSTON
HARRIS
Galveston
GALVESTON
CHAMBERS
GULF OF MEXICO
Bellville
Sealy
Rosenberg
Richmond
FORT BEND
WHARTON
Wharton
El Campo
Bay City
MATAGORDA
Matagorda B.
Freeport
Angleton
BRAZORIA
Columbus
COLORADO
La Grange
FAYETTE
Schulenburg
Flatonia
Hallettsville
LAVACA
Edna
JACKSON
Victoria
VICTORIA
Port Lavaca
CALHOUN
Matagorda I.
San Antonio B.
Austin
Lockhart
CALDWELL
Gonzales
GONZALES
Cuero
DE WITT
GOLIAD
Goliad
REFUGIO
Refugio
Aransas Pass
Rockport
ARANSAS
N.W.R.
San Marcos
HAYS
San Antonio
BEXAR
Seguin
GUADALUPE
New Braunfels
COMAL
KENDALL
Boerne
Kerrville
KERR
REAL
BANDERA
Bandera
MEDINA
Hondo
Pleasanton
ATASCOSA
KARNES
Kenedy
Beeville
BEE
LIVE OAK
George West
MCMULLEN
Tilden
Corpus Christi
NUECES
SAN PATRICIO
Sinton
Mathis
Robstown
Kingsville
KLEBERG
Corpus Christi N.A.S.
PADRE I. NAT'L SEASHORE
Padre I.
Laguna Madre
Baffin B.
KENEDY
BROOKS
Falfurrias
JIM WELLS
Alice
DUVAL
San Diego
Freer
JIM HOGG
Hebbronville
STARR
Rio Grande City
Roma
Falcon Res.
El Azúcar Res.
Sarita
Premont
BROOKS
Raymondville
WILLACY
Harlingen
San Benito
Brownsville
Matamoros
CAMERON
HIDALGO
Edinburg
McAllen
Mission
Pharr
Weslaco
Valle Hermoso
Río Bravo
Reynosa
San Fernando
Laguna Madre
Greenwich
EDWARDS
Rocksprings
VAL VERDE
Del Rio
Ciudad Acuña
KINNEY
Brackettville
LAUGHLIN A.F.B.
UVALDE
Uvalde
Sabinal
FRIO
Pearsall
Dilley
LA SALLE
Cotulla
Encinal
WEBB
Laredo
Nuevo Laredo
DIMMIT
Carrizo Springs
Crystal City
ZAVALA
Crystal City
MAVERICK
Eagle Pass
Piedras Negras
MEXICO
R í o G r a n d e
ZAPATA
Zapata
BALCONES ESCARPMENT
AMISTAD NAT'L REC. AREA
Amistad Res.
R i o G r a n d e
Nueces R.

KEY TO NUMBERS ON MAP

	County	County Seat
1	DELTA	Cooper
2	FRANKLIN	Mt. Vernon
3	CAMP	Pittsburg
4	MORRIS	Daingerfield
5	ROCKWALL	Rockwall
6	RAINS	Emory
7	GREGG	Longview
8	SOMERVELL	Glen Rose
9	SAN JACINTO	Coldspring

TERRELL VAL VERDE BIG BEND NAT'L PARK
BREWSTER
Emory Pk. 7,835'

98° West of Greenwich
100° Longitude
96°

A B C D E
5 6 7

TEXAS — 199

VERMONT

State Capitals ⊛
County Seats ⊙

MILES
0   5   10   15   20   25   30

KILOMETERS
0   5   10   15   20   25   30

© Copyright HAMMOND INCORPORATED, Maplewood, N.J.

WESTERN PART OF VIRGINIA
Same scale as main map

VIRGINIA

MILES
0 10 20 40 60

KILOMETERS

★ National Capital
⊛ State Capitals ⊙ County Seats
● Independent Cities ✶Bristol

PACIFIC OCEAN

125° 124° 122°

49

Barkley Sd.

L. Cowichan
Ladysmith
Duncan
VANCOUVER ISLAND
Port Renfrew
Victoria
Esquimalt

Strait of Juan de Fuca

C. Flattery
Neah Bay
MAKAH IND. RES.
C. Alava
Ozette
Sekiu
CLALLAM
Ozette L.
Soleduck R.
Forks
La Push
OLYMPIC
NAT'L
PARK
JEFFERSON
Mt. Olympus 7,954
OLYMPIC
NATIONAL
MTS.
PARK
Queets
Quinault
QUINAULT IND. RES.
Quinault L.
GRAYS
HARBOR
Pacific Beach
Hoquiam
Aberdeen
COAST
Central Park
Elma
Westport
Grays Har.
Grayland
Willapa B.
Leadbetter Pt.
Ocean Park
Long Beach
Nasel
C. Disappointment
WAHKIAKUM
Astoria
Columbia River

Boundary
P. B.
Pt. Roberts
Blaine Lynden
Ferndale
WHATCOM
Bellingham
Nooksack R.
Mt. Baker 10,778
Baker L.
W. Whatcom
SAN JUAN
Orcas I.
SAN JUAN N.H.P.
Anacortes
Friday Har.
Lopez I.
Fidalgo I.
Sedro-Woolley
Concrete
SKAGIT
Sooke
Oak Harbor
Whidbey I.
Burlington
Mt. Vernon
Skagit R.
Stanwood
Arlington
Darr
Admiralty
Port Townsend
Camano
Shoultes
Marysville
Lake Stevens
SNOHO
Sequim
Everett
Quilcene
Snohomish
Monroe
Mountlake Ter.
Skyko
Dabob
Lynnwood
Edmonds
Mukilteo
Bothell
Kirkland
Skykomis
Bremerton
Port Orchard
KITSAP
Burien
Renton
Maple Valley
Snoqualmie
Gig Harbor
Kent
Black Diamond
MASON
Shelton
Tacoma
Milton
Sumner
Enumclaw
Olympia
Fircrest
Parkland
Puyallup
Orting
McChord A.F.B.
LEWIS
Lacey
Tumwater
THURSTON
Tenino
Nisqually
PIERCE
MT. RAINIER
Mt. Rainier 14,410
NAT'L PK.
McCleary
Montesano
Cosmopolis
Chehalis R.
Alder L.
Ashford
Raymond
Fords Prairie
Centralia
South Bend
Pe Ell
Chehalis
LEWIS
Morton
Packwood
Long
PACIFIC
Winlock
Cowlitz
Mossyrock
R.
Toutle
Silverlake
Cathlamet
Castle Rock
Columbia Hts.
Mt. St. Helens 8,364
Mt. Adams 12,307
Longview
Kelso
COWLITZ
SKAMANIA
Kalama
Woodland
Ridgefield
Battle Ground
Stevenson
BONNEVILLE
FT. VANCOUVER NAT'L HIST. SITE
CLARK
Vancouver
Camas
Washougal
The
Portland
Mt. Hoo 11,239
Oregon City
Woodburn

48°

47°

N

46°

45°

© C.S. HAMMOND & Co., Map

INSET MAP:
122° 30' 122°
Richmond Highlands
Juanita
Redmond
Silverdale
Bainbridge
Kirkland
Bellevue
Bremerton
SEATTLE
Medina
Port Orchard
Mercer I.
L. Washington
Issaquah
Snoqualmie
Belfair
Harper
White Ctr.
Bryn Mawr
Renton
KITSAP
Normandy Pk.
Tukwila
KING
PIERCE
Des Moines
Kent
Maple Valley
Gig Harbor
Vashon
Cedar R.
Henderson B.
Tacoma
Auburn
Puget Sound
Snoqualmie R.
47° 30'
0 5 10 mi.
0 5 10 km.
122° 30' 122°

125° 124° 123° 122°
A   B   C   D

206 — WASHINGTON

WASHINGTON — 207

WEST VIRGINIA — 209

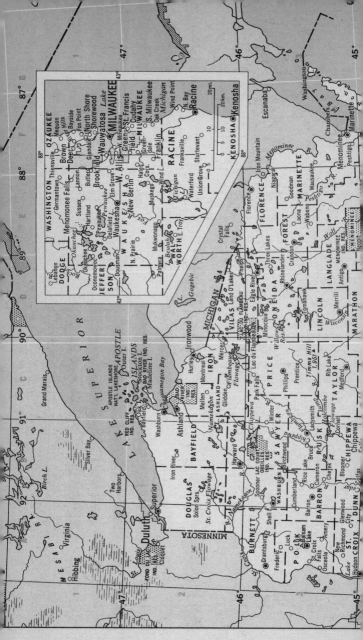

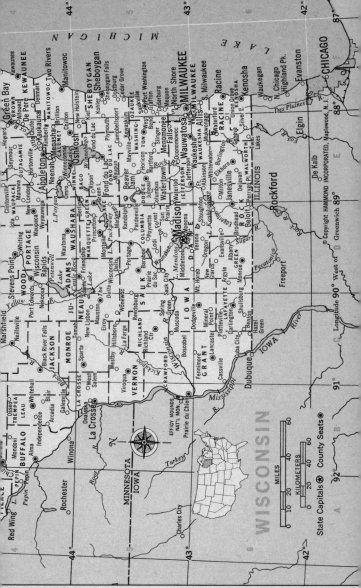

WISCONSIN — 211

WYOMING

State Capital ⊛

MILES
0 10 20 40 60

KILOMETRES
0 20 40 60

106° D 104° E

44°

Hulett
DEVILS TOWER NAT'L MON.
CROOK
BEAR LODGE MTS.
● Sundance

Ucross
Warmont

CAMPBELL
Gillette ●
Keyhole Res.
● Moorcroft

Buffalo

BLACK

Rapid City ○

NSON

Belle Fourche R.
Powder R.

● Upton
● Osage
WESTON
Newcastle ●
HILLS

MT. RUSHMORE NAT'L MEM. ■
Custer ○
JEWEL CAVE NAT'L MON. ■
WIND CAVE NAT'L PK.

Linch
Antelope Cr.
Cheyenne

Hot Sprs. ○

PINE RIDGE IND. RES.

Midwest
Edgerton ●
Teapot Dome ●

Angostura Res.

NIOBRARA

SOUTH DAKOTA
NEBRASKA

ONA

CONVERSE
Lance Creek ●

Mountain View ●
Mills ● Evansville ●
Casper ●
Glenrock ●
Douglas ●
Orin ●
Paradise Valley ●

Manville ●
● Lusk

Chadron ○

Niobrara R.

Alcova Res.
inder

Glendo Res.
● Glendo
Guernsey Res.
● Sunrise

AGATE FOSSIL BEDS NAT'L MON. ■

N

LARAMIE

Laramie Pk. ▲
● Guernsey
FT. LARAMIE NAT'L HIST. SITE
● Ft. Laramie
● Lingle

North Platte River

inoe

● Wheatland
Torrington ●
GOSHEN

42°

Medicine Bow ●
Hanna ●
Wheatland Res.
● Rock River
PLATTE
● Chugwater

● Hawk Sprs.

Scottsbluff ○
SCOTTS BLUFF NAT'L MON. ■

River

Elk Mtn. ▲
● McFadden

CHIMNEY ROCKS NAT'L HIST. SITE

ICINE BOW RANGE

ALBANY

LARAMIE

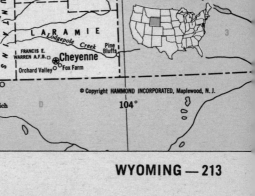

toga

Laramie ●
Lodgepole Creek
● Pine Bluffs

MOUNTAINS

FRANCIS E. WARREN A.F.B. ●
Cheyenne ⊛

Platte R.
Laramie R.

● Foxpark
● Orchard Valley ● Fox Farm

ent

COLORADO
© Copyright HAMMOND INCORPORATED, Maplewood, N.J.

Longitude West 106° of Greenwich D 104°

WYOMING — 213

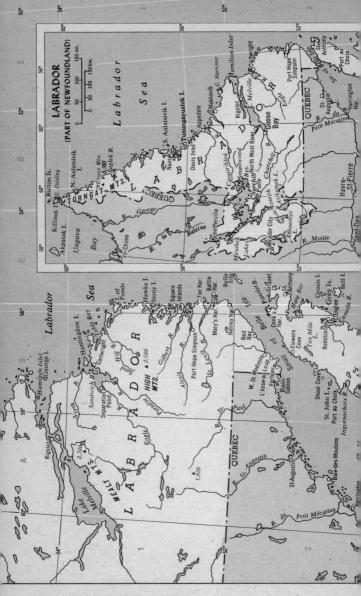

LABRADOR
(PART OF NEWFOUNDLAND)

150 mi.
50 100 150 km.

Labrador Sea

QUEBEC

Button Is.
Killinek I.   C. Chidley
Akpatok I.   N. Aulatsivik I.
Ungava Bay   Torngat Mtns. ▲5,160
TORNGAT MTS.   Nulvek B.
                S. Aulatsivik I.
George   Nain
R. la   Davis Inlet
Baleine   Hopedale
Schefferville   Utunungayualok I.
                Makkovik
                C. Harrison
                Hamilton Inlet
                Cartwright
                Port Hope Simpson
Menihek   Rigolet   Melville
Lake   Camairiktok   Goose Bay
        Smallwood   Eske
Labrador City   Res.   L A B R A D O R
Joseph L.   North West River
Ashuanipi   Churchill   Churchill
Attikonak L.   Falls
R. Moisie   Ashuanipi

St. Anthony
Port au Choix
QUEBEC
St. Augustin
R. du
Petit Mécatina
Havre-St-Pierre
Sept-Îles
St-Pierre

Labrador Sea

Hamilton Inlet   I. of Ponds   Hawke I.
George L.   Stony I.
Huntingdon I.   Square Islands
Double Bay   Fox Har.
Cartwright   Battle Har.
Hill   Belle Isle
HIGH MTS. ▲2,080   Mary's Har.
Gilbert R.   Henley Har.   Cook's Har.
Sandwich B.   Red   Griquet
Separation Point   Bay   St. Anthony
North   Alexis   W. St. Modeste   Flowers   Hare Bay
        Forteau   Cove   Groais I.
3,700   Paradise   L'Anse-au-Loup   Roddickton   Grey Is.
MEALY MTS.   Blanc-   Englee   Canada B.
Eagle   Sablon   Ten Mile
Rigolet   Shoal Cove
1,920   St. John I.   Hr.
Lake   St.   Port au Choix
Melville   QUEBEC   Paul   Ingrnachoix B.
        R. St. Augustin   St-Augustin
        Baie-des-Moutons
        R. du Petit Mécatina

50° 54°
52°
56°
60°
64°
68°

214 — NEWFOUNDLAND

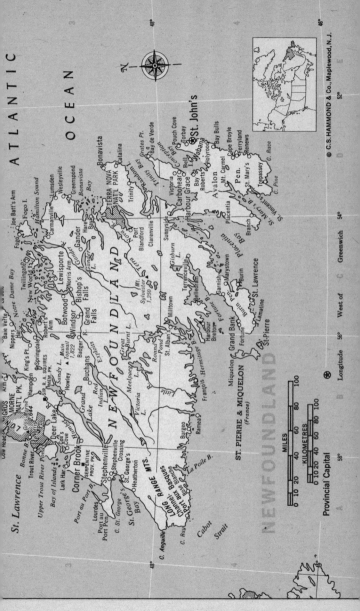

© C.S. HAMMOND & Co., Maplewood, N.J.

NEWFOUNDLAND — 215

Gulf of

St. Lawrence

Magdalen Is. *(Que.)*

St. Paul I.

C. North

Cape North

Aspy Bay

CAPE BRETON
HIGHLANDS
NAT'L PK. *1,747*

Petit-Étang

Chéticamp

Ingonish

Grand-Étang

CAPE

Margaree

St. Ann's B.

Bolarderie I.

Florence

Sydney Mines

Margaree Forks

Goose Cove

New Waterford

Inverness

BRETON N

Sydney

Glace Bay

L. Ainslie

Baddeck

Dominion

Mabou

Sydney

Reserve Mines

Donkin

Port Hood

1,075

ISLAND

Mira R.

Scatarie I.

Bras d'Or
Lake

Louisbourg

Mira

C. Breton

LOUISBOURG N.H.P.

C. George

Georges
Bay

Gabarus

Gabarus B.

Strait

Pictou I.

Port
Hawkesbury

St. Peters

Pictou

Trenton

Antigonish

Louisdale

New Glasgow

Thorburn

Mulgrave

Canso

Arichat

I. Madame

Westville

Stellarton

Boylston

Hopewell

Guysborough

Chedabucto B.

Upper Stewiacke

Canso

Andrew I.

Sherbrooke

Goldboro

Tor
B.

Ecum
Secum

le Musquodoboit

Country Har.

Sheet Har.

Moser R.

Ship Har.

dit

ATLANTIC OCEAN

'WARD ISLAND
PARK

St. Peters Bay

East Pt.

'D ISLAND

Souris

Georgetown

Montague

Cardigan B.

C. Bear

West Pt.

Sable I.

NOVA SCOTIA

MILES

0 10   20   40   60

KILOMETRES

0   20   40   60

Provincial Capitals

⊛

Longitude 62° West of Greenwich 61°

MILES
0 10 20 40 60

KILOMETRES
0 10 20 40 60

Provincial Capitals

C. S. HAMMOND & Co., Maplewood, N.J.

NEW BRUNSWICK

218 — NEW BRUNSWICK

220 — PRINCE EDWARD ISLAND

PRINCE EDWARD
ISLAND

MILES
0 5 10 15 20

KILOMETRES
0 5 10 15 20

Provincial Capital ✪

47°

63° 62° 30' 62°

47°

_O_ _F_

_L A W R E N C E_

N

46°
30'

Savage
Har.
Monticello

St. Peters
Bay
Baltic Elmira
East
Pt.

tanhope Morell Bear River New Zealand Kingsborough
Tracadie St. Peters
Bay St. Charles Souris
d Grand Tracadie Bangor Fortune Br.
Mt. Stewart Strathcona Rollo Bay

Hillsborough Ft. Augustus Peakes Little Pond
Id (Peakes Rd.) Spry Pt.
erwoods Lorne Valley Boughton
Parkdale Bay
Bunbury L. Verde Cardigan Bruce Pt.
New Perth Newport Boughton I.
Crossroads Millview Cardigan R.
Cherry Valley Vernon Bridge Montague Georgetown
borough Lower Cardigan Bay
ers ▽ Montague Panmure I.
Bay Orwell B. Valleyfield Sturgeon
t. Prim Belfast Iona Peters Road Murray Har. N.
Pinette R. 450 ▲ Murray Har.
Murray Beach Pt.
River C. Bear
Belle River Murray R. Abney Murray
Wood Har.
Islands
Wood Is.

_S T R A I T_

NOVA SCOTIA

© C.S. HAMMOND & Co., Maplewood, N.J.

vich 63° D 62° 30' 62°

46°

QUÉBEC

0 100 200 mi.

0 100 200 km.

N.W.T.

Mansel I.
Saglouc
Baffin I.
Hudson
Str.
Nouveau-Québec
Maricourt
Crater
Povungnituk
Ungava
Pen.
Ungava
B.
Port-Nouveau-Québec
L. Payne
R. aux Feuilles
Ft-Chimo
Nain
ATLANTIC
OCEAN
Hudson
Bay
Belcher
Is.
Lac à l'Eau Claire
Schefferville
Hopedale
LABRADOR
Poste-de-
la-Baleine
L. Bienville
La Grande-Rivière
Smallwood
Res.
Labrador
City
James
B.
Ft-George
Akimiski
R. Eastmain
Rupert House
NEWF.
La Sarre
Mistassini
Chibougamau
Sept-Iles
Havre-St-Pierre
Noranda
Rouyn
Val-d'Or
LAURENTIDES
PROV. PK.
Chicoutimi
Rimouski
Ile d'Anticosti
NEWF.
Gulf of
St. Lawrence
Gaspé
Québec
St. Lawrence
Mt-Laurier
Fredericton
N.B.
P.E.I.
Magdalen Is.
Charlottetown
Sydney
Ottawa
MONTRÉAL
N.Y.
ME.
N.S.
Halifax
ATLANTIC
OCEAN

Mistas
Normandin
Dolb
St-Féli
Rober
L. des
Commissaire
R. Trenche
R. Croche
Rapide-Blanc
R. Windigo
R. Vermilion
L.
Blanc
La Tuque
L.
Edouard
R. St. Maurice
L.
Wayagama
LA MAURICIE
NAT'L PARK
Ste-Thècle
Res. Matawin
Ste-Tite
La Pér
MONT-TREMBLANT
PROV.
PARK
Grand'Mère
Shawinigan
Cap-
Made
Baskatong
Ferme-
Neuve
St-Michel-des-Saints
Mont-Laurier
St-Alexis-
des-Monts
Trois-Rivières
Maniwaki
L'Annonciation
Mt.
Tremblant
3,150
St-Donat-
de-Montcalm
St-Gabriel
St-Félix-de-Valois
Nicolet
St-Pierre
Pierreville
L. du Poisson-
Blanc
L.
Gagnon
Labelle
St-Jovite
R. L'Assomption
R. Rouge
Ste-Agathe
Val-David
Berthierville
Joliette
St-Jacques
Tracy
Sorel
Drummondville
V.
L.
Simon
Ste-Adèle
L'Épiphanie
L'Assomption
Contrecœur
Verchères
GATINEAU
PARK
Buckingham
Thurso
Montebello
St-Sauveur
St-Jérôme
Lachute
Ste-Thérèse
Pte-aux-Trembles
St-Hyacinthe
Ric
Acton Va.
Hull
Gatineau
Hawkes-
bury
R.
Rigaud
Laval
Lachine
St-Laurent
Longueuil
Verdun
La Providence
Bre
Sh
Aylmer
Ottawa
Alexandria
Dorion
MONTRÉAL
Marieville
Granby
Waterlo
MT-
PRO
M
Carleton
Place
Kemptville
Beauharnois
Lac Ste-
Claire
St-Jean
Iberville
Cowansville
Farnham
Lac-Brome
Smiths Falls
Cornwall
St. Lawrence R.
Valleyfield
Ormstown
Huntingdon
St-Rémi
Napierville
Bedford
Sutton
L.
Memphremagos
Newpo
Massena
NEW YORK
Rouses Pt.
L. Champlain
73° Longitude
W

E 71° F 70° G 69° H 68°

Péribonca

Canton Bégin
St-Coeur-de-Marie
Kénogami Chicoutimi R. Saguenay
ma Jonquière
biens Arvida Bagot- R. Portneuf Forestville RIVER
ortville L. Kénogami ville Port-Alfred Mont-Joli Price
Luceville
Tadoussac Bic
Île Trois-Pistoles Rimouski
Verte St-Fabien

LAURENTIDES I. aux L'Isle-Verte 48°
Lièvres NEW
PROVINCIAL St-Siméon Rivière-du-Loup BRUNSWICK
▲ 3,925 Clermont L. Temiscouata
PARK La Malbaie Cabano
Pte-au-Pic Notre-Dame-du-Lac Dégelis N.B. 2
Baie-St-Paul St-Pascal St-Joseph-de- Edmundston
I. aux St-Pacôme St-Éleuthère la-Rivière-Bleue
Coudres La
St-Jean- Pocatière QUEBEC
Port-Joli
Ste-Anne- I. aux L'Islet MILES
de-Beaupré Oies Cap-St-Ignace 0 10 20 40 60
Montmorency Île St-Pamphile
Charlesbourg d'Orléans St-Pierre- KILOMETRES
Québec Lévis Montmagny 0 20 40 60
St-Foy St-Raphaël
Donnacona St- Lac-Frontière National Capital
ux Romuald St-Gervais Provincial Capitals
St-Agapitville d'Etchemin
St-Anselme Ste-Claire-de-Joliette
Plessisville Ste-Marie Lac-Etchemin
rince- Tring-Jct. St-Joseph-de-Beauce
ville Bernierville Beauceville-Est
nabaska Black L. St-Georges St-Zacharie
Disraëli Linière
k Thetford Mines
L. Aylmer Bolduc 66° R. 64°
estos St-François La Lawrence Grande-Vallée
Weedon-Centre Guadeloupe St. Cap-Chat Ste-Anne-des-Monts Mt. Jacques Cartier
E-Angus Scotstown Lac- Les GASPÉSIE 4,160 Murdochville FORILLON
Cookshire Mégantic 68° Méchins PROV. NAT'L PARK
Matane PK. C. du
nnoxville N.H. MAINE 48° Price Cascapédia Gaspé Gaspé
terville Mont- Amqui Gaspé Peninsula Grande-Rivière Percé
Joli Sayabec Chandler
oaticook Causapscal Bonaventure Gulf of
nd Matapédia New St. Lawrence
ONT NEW BRUNSWICK Campbellton Carlisle 0 20 40 mi.
0 20 40 km. 45°
68° 66° 64°
© C.S. HAMMOND & CO., Maplewood, N.J.

E 71° F 70° G 69° H 69°

QUEBEC — 223

ONTARIO

MILES
0 10 20 40 60 80

KILOMETERS
0 20 40 60 80

● National Capital
● Provincial Capital

QUÉBEC

Deep River
Chalk River
Petawawa

Ft-Coulonge

GATINEAU PARK

Hawkesbury
Rockland   Vankleek Hill
Hull   Vanier
                    Dorion
Casselman   Alexandria

Ottawa

QUIN PROVINCIAL
PARK

Whitney
Killaloe Sta.
Mada-
waska
Pembroke

Cobden

Eganville   Renfrew

Barry's Bay

Madawaska

Arnprior
Stittsville
Richmond

Almonte
Carleton Place

Chesterville
Winchester
Kemptville
Morrisburg
Iroquois
Cardinal
Prescott

Cornwall

N. Y.

Massena

Malone

Haliburton
Bancroft   Weslemkoon
L.
Mississippi L.

Mazinau L.

Smiths Falls
Sharbot L.
Perth

Ogdensburg

Rideau

Rideau R.

inden

enelon Falls
bbcaygeon
ha Lakes
Lakefield
Marmora
Madoc
Havelock
Tweed
Norwood
Campbellford
Napanee

Lakes
Brockville

ST. LAWRENCE IS.
NAT'L PARK
Gananoque

Rice
L.
Frankford
Trenton
Belleville
Deseronto

Kingston

St. Lawrence

Wolfe I.

Brighton
Picton
Prince
Edward Pen.

Amherst
I.

castle
awa
Port Hope
Colborne
PT. PETRE

ONTARIO

ort
Brockport

Falls

ffalo   Batavia
NEW YORK
Warsaw

ITED

ATES

alamanca
Olean

heny R.

HUDSON BAY

MANITOBA

Belcher Is.

QUÉBEC

Ft-George

POLAR
BEAR
PROV.
PARK

Severn R.

Trout Lake

Winisk R.

James

Akimiski I.

Sakami
L.

Island
L.

Favourable Lake

Winnipeg R.

Red Lakes
L. St.
Joseph
Pickle
Crow

Attawapiskat

Albany R.

Ft. Albany

Moosonee

Bay

Rupert House

Kenora
Lac
Seul
Sioux
Lookout
Armstrong
Auden

Nipigon L.

Geraldton
Hearst
Manitouwadge

Fraserdale
Kapuskasing
Cochrane

Kirkland
Lake

Lake of
the
Woods
Dryden
Atikokan
QUETICO PROV. PK.

Ft. Frances

Nipigon

Schreiber

Marathon

Wawa

Timmins

New
Liskeard
Haileybury

North
Bay

Thunder
Bay

PUKASKWA
NAT'L PARK

SUPERIOR
PROV. PARK

Chapleau

Levack

ONTARIO
NORTHERN PART

0   100   200 mi.

0   100   200 km.

Lake
Superior

Sault
Ste. Marie

MICH.

Marquette

Sudbury

QUÉBEC

© C.S. HAMMOND & Co., Maplewood, N.J.

Longitude 78° West of Greenwich

ONTARIO — 225

© C.S. HAMMOND & Co., Maplewood, N.J.

MANITOBA — 227

© C.S. HAMMOND & Co., Maplewood, N.J.

MONTANA

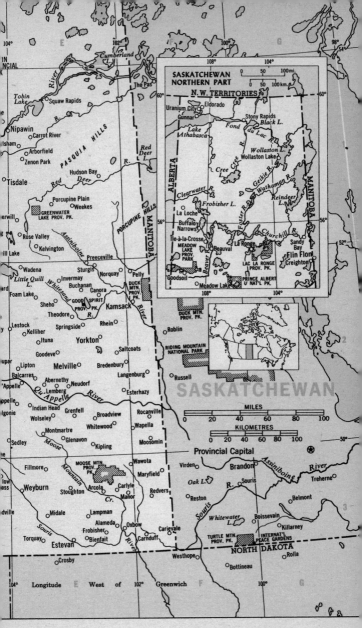

SASKATCHEWAN — 229

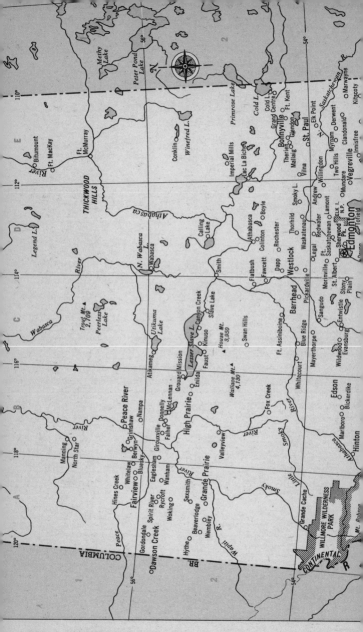

230 — ALBERTA

ALBERTA — 231

BRITISH COLUMBIA — 233

234 — YUKON TERRITORY

YUKON TERRITORY — 235

NORTHWEST TERRITORIES

MILES
0 50 100 200 300 400

KILOMETRES
0 50 100 200 300 400

State and Territorial Capitals ✪

ARCTIC OCEAN

Beaufort Sea

ELIZ

ISLANDS

Borden I.

Prince Patrick I.

Macken King I.

Lands End

Mould Bay

Hazen

PARRY

Melville I.

C. Pr. Alfred

M'Clure Str.

Viscount Melvi

Sound

Parry

Sachs Har.

Banks I.

C. Lambton

Amundsen Gulf

Hadle Bay

DISTRICT

Victoria

BROOKS R.A.

UNITED STATES

Barrow

Pt. Barrow

Mackenzie

Herschel

Bay

Tuktoyaktuk

Holman I.

Island

Yukon

8,500

Old Crow

Aklavik

C. Bathurst

Inuvik

Paulatuk

Wollaston Pen. ▲1,700

Ca

Fairbanks

ALASKA

Yukon River

Ft. McPherson

Ft. Good Hope

Bluenose L.

Coppermine

Caronation Gulf

Arctic Circ

▲7,400

Dawson

Mayo

Elsa

YUKON TERR.

Mackenzie

Norman Wells

Keele Pk. 9,750

Ft. Franklin

Great Bear Lake

Port Radium

Contwoyto L.

Beaver Cr.

Pelly

Carmacks

Kluane

Ross R.R.

Macmillan

Ft. Norman

DISTRICT

Thelo

▲8,000 Mt. Sir James MacBrien

Destruction Bay

Mt. Logan KLUANE NAT'L PARK

Teslin

Whitehorse

NAHANNI NAT'L PARK

Wrigley

Mackenzie River

OF MACKENZI

GLACIER BAY N.M.

Juneau

BRITISH

Watson L.

Lac la Martre

Lac la Martre

Rae-Edzo

Yellowknife

ALASKA

Skagway

COAST MTS.

COLUMBIA

Liard

ROCKY

Ft. Simpson

R.

Ft. Providence

Great Slave Lake

Snowdrift

Nonacho

Petersburg

Ft. Liard

MUNCHO L. PROV. PK.

MTS.

Ft. Nelson

Hay River

Ft. Resolution

Ft. Smith

Wholdaia

ALBERTA

WOOD BUFFALO NAT'L PK.

Pine Pt.

SASKATCH

Meander R.

Longitude

120°

Uranium City

West

236 — NORTHWEST TERRITORIES

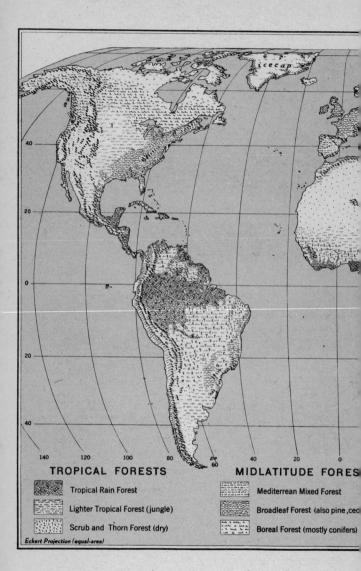

icecap

40

20

0

20

40

140 120 100 80 60 40 20 0

TROPICAL FORESTS

Tropical Rain Forest

Lighter Tropical Forest (jungle)

Scrub and Thorn Forest (dry)

Eckert Projection (equal-area)

MIDLATITUDE FORES

Mediterrean Mixed Forest

Broadleaf Forest (also pine, cec

Boreal Forest (mostly conifers)

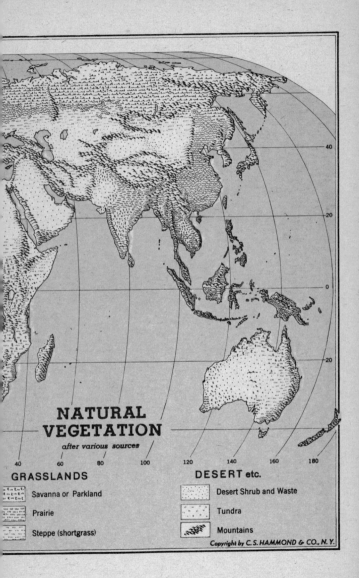

NATURAL VEGETATION

after various sources

| | 40 | 60 | 80 | 100 | | 120 | 140 | 160 | 180 |

GRASSLANDS

Savanna or Parkland

Prairie

Steppe (shortgrass)

DESERT etc.

Desert Shrub and Waste

Tundra

Mountains

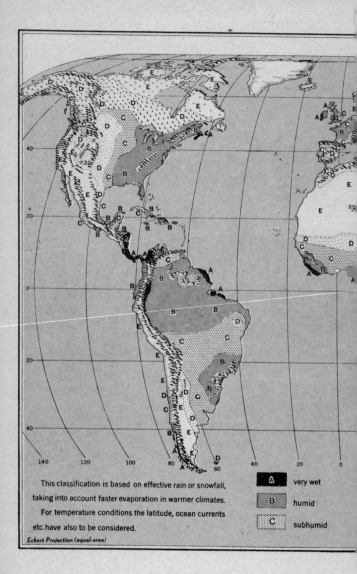

This classification is based on effective rain or snowfall, taking into account faster evaporation in warmer climates. For temperature conditions the latitude, ocean currents etc. have also to be considered.

Eckert Projection (equal-area)

A	very wet
B	humid
C	subhumid

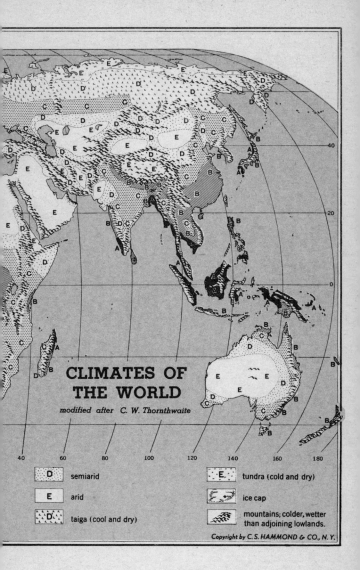

CLIMATES OF
THE WORLD

modified after C. W. Thornthwaite

D	semiarid	E	tundra (cold and dry)
E	arid		ice cap
D	taiga (cool and dry)		mountains; colder, wetter than adjoining lowlands.

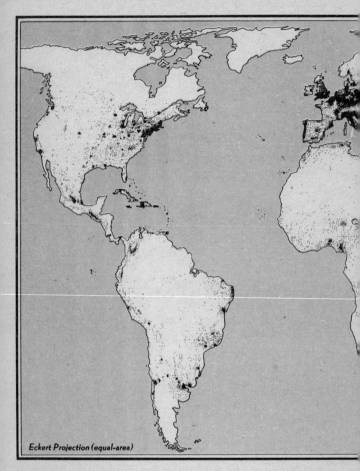

Eckert Projection (equal-area)

DENSITY OF POPULATION. One of the most outstanding facts
human geography is the extremely uneven distribution of people over
Earth. One-half of the Earth's surface has less than 3 people per square m
while in the lowlands of India, China, Java and Japan rural density reac

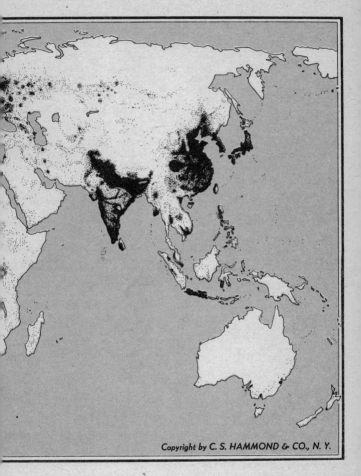
Copyright by C. S. HAMMOND & CO., N. Y.

incredible congestion of 2000-3000 per square mile. Three-fourths of the
rth's population live in four relatively small areas; Northeastern United
tes, North-Central Europe, India and the Far East.

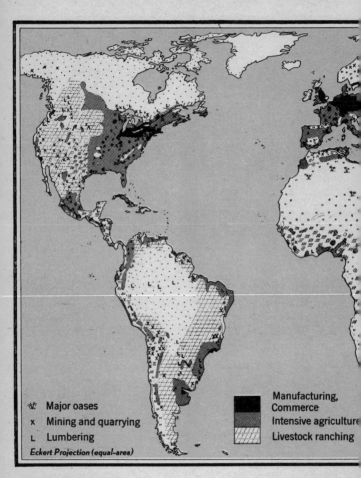

Correlation with the density of population shows *the most densely populated areas fall into the regions of manufacturing* *intensive farming. All other economies require considerable space. The m*

Major oases
x   Mining and quarrying
L   Lumbering
Eckert Projection (equal-area)

Manufacturing, Commerce
Intensive agriculture
Livestock ranching

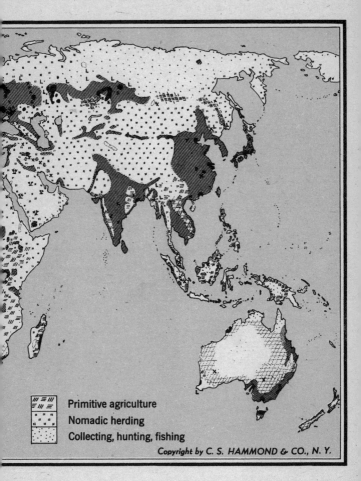

	Primitive agriculture
	Nomadic herding
	Collecting, hunting, fishing

Copyright by C. S. HAMMOND & CO., N. Y.

parsely inhabited areas are those of collecting, hunting and fishing. Areas ith practically no habitation are left blank.

LANGUAGES. *Several hundred different languages are spoken in the World, and in many places two or more languages are spoken, sometimes by the same people. The map above shows the dominant languages in each*

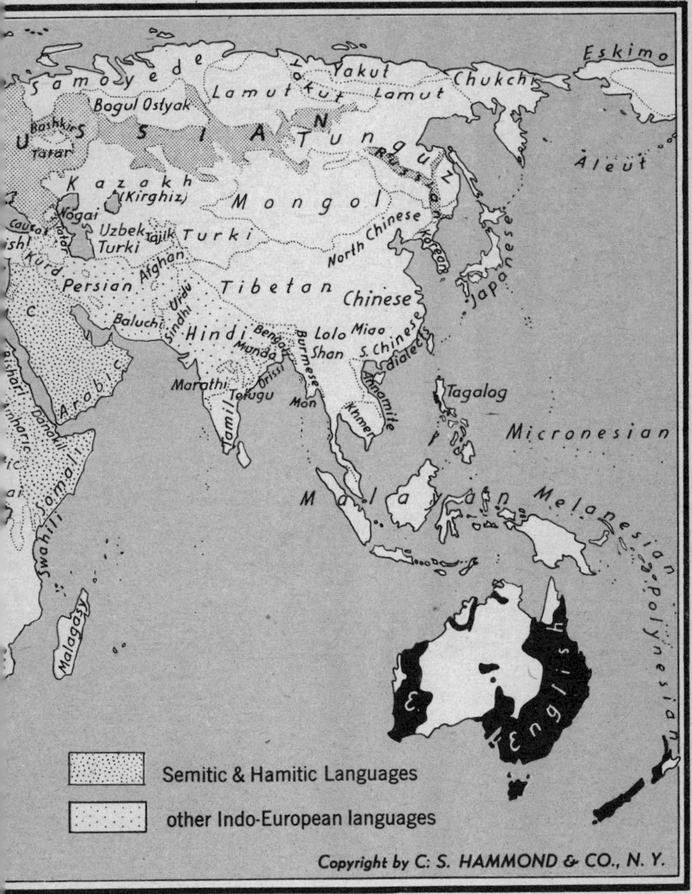

Samoyede Yakut Chukch Eskimo

Bashkirs Bogul Ostyak Lamut Lamut Aleut
Tatari

U S S I A Tunguz

Kazakh (Kirghiz) Mongol
Nogai
Caucas- Uzbek Tajik Turki North Chinese Japanese
ish! Turki
Kurd Persian Afghan Chinese
Urdu Tibetan
Baluchi Sindhi Hindi Lolo Miao
Munda Bengali S. Chinese
Arabic Marathi Burmese Shan
Telugu Mon Annamite Tagalog
Tamil Khmer Micronesian

Somali Malay an Melanesian
Swahili Polynesia

Malagasy English

℈

English

ality. English, French, Spanish, Russian, Arabic and Swahili are spoken by
ny people as a second language for commerce or travel.

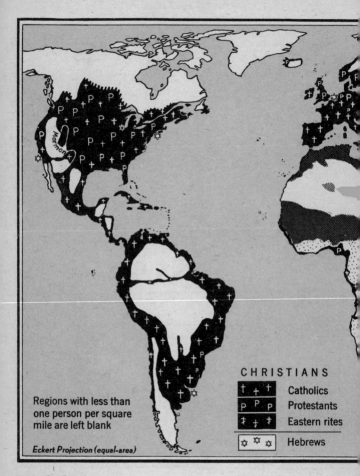

CHRISTIANS

† † †	Catholics
P P P	Protestants
‡ † ‡	Eastern rites
✡ ✡ ✡	Hebrews

Regions with less than
one person per square
mile are left blank

Eckert Projection (equal-area)

RELIGIONS. Most people of the Earth belong to four major religio
Christians, Mohammedans, Brahmans, Buddhists and derivatives. The East
rites of the Christians include the Greek Orthodox, Greek Catholic, Armeni
Syrian, Coptic and more minor churches. The lamaism of Tibet and Mongo

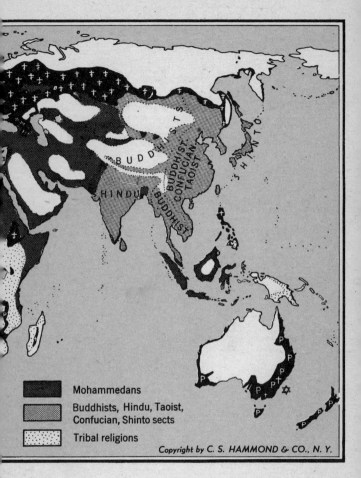

Mohammedans

Buddhists, Hindu, Taoist,
Confucian, Shinto sects

Tribal religions

Copyright by C. S. HAMMOND & CO., N. Y.

fers a great deal from Buddhism in Burma and Thailand. In the religion of ina the teachings of Buddha, Confucius and Tao are mixed, while in Shinto great deal of ancestor and emperor worship is added. About 11 million brews live scattered over the globe, chiefly in cities and in the state of ael.

INDEX OF THE WORLD

This alphabetical list gives statistics of population based on the latest official reports. Each line begins with the name of a place or feature, followed by the name of the country or state, the population, the index reference and the page number.

Capitals are designated by asterisks*   † Including suburbs

Salem, India, 515,021 ....C 3 64
Salem,* Oreg., 89,223 ....A 2 186
Salerno, Italy, 146,534 ....E 4 43
Salford, Eng., 261,000 ....G 2 30
Salisbury,* Zimbabwe, †601,000 ...G 5 83
Salonika (Thessaloníki), Greece,
  345,799 ....C 3 49
Salta, Argentina, 176,216 ....C 2 98
Saltillo, Mexico, 200,712 ....E 3 102
Salt Lake City,* Utah, 163,033 ...C 3 200
Salto, Uruguay, 72,948 ....E 4 98
Salton Sea (lake), Calif. ....E 6 123
Salvador, Brazil, 1,525,831 ....E 5 97
Salween (river), Asia ....B 2 66
Salzburg, Austria, 139,000 ....B 3 46
Salzgitter, Ger., 113,500 ....D 2 34
Samaria (reg.), Jordan ....C 3 61
Samarinda, Indonesia, 137,521 ...F 6 74
Samarkand, U.S.S.R., 477,000 ...C 6 52
Samsun, Turkey, 150,947 ....G 2 59
San'a,* Yemen, 134,588 ....E 6 56
San Angelo, Tex., 73,240 ....B 4 198
San Antonio, Tex., 785,410 ....C 5 199
San Bernardino, Calif., 118,057 ...E 5 123
San Cristóbal, Venez., 152,239 ...C 2 94
San Diego, Calif., 875,504 ....E 6 123
Sandy Hook (spit), N.J. ....E 3 172
San Francisco, Calif., 678,974 ...D 2 122
Sangre de Cristo (mts.), U.S. ...E 3 124
San Joaquin (riv.), Calif. ....C 4 123
San Jose, Calif., 636,550 ....F 2 122
San José,* Costa Rica, 215,441 ...D 3 105
San Juan, Argentina, 112,582 ...C 4 98
San Juan,* Puerto Rico, 422,701 .G 1 107
Sankt Gallen, Switz., 81,900 ....B 1 42
San Leandro, Calif., 63,952 ....E 2 122
San Luis Potosí, Mex., 281,534 ...E 3 103
San Mateo, Calif., 77,561 ....E 2 122
San Pedro Sula, Hond., 213,600 .B 2 104
San Salvador,* El Salvador,
  368,000 ....B 2 104
San Salvador (isl.), Bahamas ....D 2 106
San Sebastián, Spain, 165,829 ...E 1 41
Santa Ana, Calif., 203,713 ....F 3 122
Santa Ana, El Salvador, 96,306 ..B 2 104
Santa Barbara, Calif., 74,542 ....C 5 123
Santa Clara, Calif., 87,746 ....E 2 122
Santa Clara, Cuba, 150,000 ....B 2 106
Santa Cruz, Bolivia, 254,682 ....E 7 95
Santa Cruz de Tenerife,* Canary
  Islands, Spain, 151,361 ....B 4 40
Santa Fe, Argentina, 244,655 ....D 4 98
Santa Fe,* N. Mex., 48,899 ....B 2 174
Santa Maria, Brazil, 120,667 ....F 4 98
Santa Marta, Colombia, 102,484 .C 1 94
Santa Monica, Calif., 88,314 ....D 3 122
Santander, Spain, 149,704 ....D 1 40
Santiago,* Chile, 3,691,548 ....B 4 98
Santiago, Dom. Rep., †173,975 ...D 3 106
Santiago, Spain, 70,893 ....B 1 40
Santiago de Cuba, Cuba, 322,000 .C 3 106
Santiago del Estero, Arg.,
  105,127 ....D 3 98

Santo Domingo,* Dom. Rep.,
  †802,619 ....E 3 107
Santos, Brazil, 341,317 ....C 7 97
São Francisco (river), Brazil ....D 4 97
São Luís, Brazil, 460,320 ....D 3 96
Saône (river), France ....F 4 39
São Paulo, Brazil, 8,584,896 ....C 7 97
Sapporo, Japan, 1,400,000 ....F 2 71
Saragossa, Spain, 542,317 ....F 2 41
Sarajevo, Yugoslavia, †400,000 ..B 3 48
Saransk, U.S.S.R., 263,000 ....F 4 51
Saratov, U.S.S.R., 856,000 ....F 4 51
Sarawak (state), Malaysia ....E 5 74
Sardinia (isl.) Italy ....B 4 43
Sargodha, Pakistan, 201,407 ....C 1 64
Sarnia, Ontario, 55,576 ....A 3 224
Sasebo, Japan, 250,723 ....C 4 70
Saskatchewan (riv.), Canada ....D 1 228
Saskatoon, Sask., 141,600 ....C 1 228
Sassari, Italy, 94,312 ....B 4 43
Satu Mare, Romania, 103,612 ....C 2 48
Sault-Sainte-Marie, Ont., 81,048 .D 3 225
Savannah, Ga., 141,634 ....E 4 133
Savona, Italy, 76,274 ....B 2 42
Schenectady, N.Y., 67,972 ....F 3 177
Schiedam, Netherlands, 78,068 ...D 4 36
Schwerin, Germany, 117,406 ....D 2 34
Scottsdale, Ariz., 88,364 ....C 3 119
Scranton, Pa., 88,117 ....F 2 189
Seattle, Wash., 493,846 ....B 4 206
Secunderabad, India, 250,836 ....C 3 64
Seine (river), France ....D 3 38
Semarang, Indonesia, 646,590 ....J 2 75
Semipalatinsk, U.S.S.R., 283,000 .D 4 52
Sendai, Japan, 615,473 ....F 3 71
Seoul,* S. Korea, 8,366,756 ....B 3 70
Sept-Îles, Quebec, 30,617 ....B 2 222
Seremban, Malaysia, 91,130 ....C 7 67
Serov, U.S.S.R., 101,000 ....K 3 50
Serpukhov, U.S.S.R., 140,000 ....E 4 51
Sétif, Algeria, 88,212 ....F 1 78
Sevastopol, U.S.S.R., 301,000 ....D 6 51
Severn (riv.), U.K. ....E 5 31
Severnaya Zemlya (isls.),
  U.S.S.R. ....G 1 52
Severodvinsk, U.S.S.R., 197,000 ..E 2 50
Seville, Spain, 588,784 ....C 4 40
Sfax, Tunisia, †256,739 ....G 1 78
'sGravenhage (The Hague),
  *Neth., 467,997 ....C 3 36
Shahjahanpur, India, 205,325 ....D 2 64
Shakhty, U.S.S.R., 209,000 ....F 5 51
Shanghai, China, 11,320,000 ....F 2 69
Shannon (river), Ireland ....A 4 31
Shantou, China, 400,000 ....D 3 69
Shaoxing, China, 225,000 ....E 3 69
Shaoyang, China, 275,000 ....E 3 69
Shasta (mt.), Calif. ....C 2 122
Shawinigan, Quebec, 24,921 ....C 3 222
Sheffield, England, 536,800 ....F 4 31
Shenyang, China, 2,690,000 ....F 1 69
Sherbrooke, Quebec, 76,804 ....D 4 222
'sHertogenbosch, Neth., 85,511 ...G 4 36